PHYSICS
for the IB Diploma
STANDARD AND HIGHER LEVEL

Tim Kirk

OXFORD

UNIVERSITY PRESS

Great Clarendon Street, Oxford OX2 6DP

Oxford University Press is a department of the University of Oxford.
It furthers the University's objective of excellence in research,
scholarship, and education by publishing worldwide in

Oxford New York

Auckland Cape Town Dar es Salaam Hong Kong Karachi
Kuala Lumpur Madrid Melbourne Mexico City Nairobi
New Delhi Shanghai Taipei Toronto

With offices in

Argentina Austria Brazil Chile Czech Republic France Greece
Guatemala Hungary Italy Japan Poland Portugal Singapore
South Korea Switzerland Thailand Turkey Ukraine Vietnam

Database right Oxford University Press (maker)

First published 2003

British Library Cataloguing in Publication Data

Data available

ISBN 13 9780199148363

ISBN 0 19 914836 8

10 9 8 7 6 5

Design by Linda Reed & Associates, Shaftesbury, Dorset

Illustration by Barking Dog Art, Nailsworth, Gloucestershire

Cover photo: Getty Images

Printed in Great Britain

Introduction and acknowledgements

Many people seem to think that you have to be really clever to understand Physics and this puts some people off studying it in the first place. So do you really need a brain the size of a planet in order to cope with IB Higher Level Physics? The answer, you will be pleased to hear, is 'No'. In fact, it is one of the world's best kept secrets that Physics is easy! There is very little to learn by heart and even ideas that seem really difficult when you first meet them can end up being obvious by the end of a course of study. But if this is the case why do so many people seem to think that Physics is really hard?

I think the main reason is that there are no 'safety nets' or 'short cuts' to understanding Physics principles. You won't get far if you just learn laws by memorising them and try to plug numbers into equations in the hope of getting the right answer. To really make progress you need to be familiar with a concept and be completely happy that you understand it. This will mean that you are able to apply your understanding in unfamiliar situations. The hardest thing, however, is often not the learning or the understanding of new ideas but the getting rid of wrong and confused 'every day explanations'.

This book should prove useful to anyone following a pre-university Physics course but its structure sticks very closely to the recently revised International Baccalaureate syllabus. It aims to provide an explanation (albeit very brief) of all of the core ideas that are needed throughout the whole IB Physics course. To this end each of the sections is clearly marked as either being appropriate for everybody or only being needed by those studying at Higher level. The same is true of the questions that can be found at the end of the chapters.

I would like to take the opportunity to thank the many people that have helped and encouraged me during the writing of this book. In particular I need to mention David Jones from the IB curriculum and assessment offices in Cardiff and Paul Ruth who provided many useful and detailed suggestions for improvement – unfortunately there was not enough space to include everything. The biggest thanks, however, need to go to Betsan for her support, patience and encouragement throughout the whole project.

Tim Kirk
October 2002

CONTENTS

The realm of Physics – Range of magnitudes of quantities in our universe

ORDERS OF MAGNITUDE – INCLUDING THEIR RATIOS

As stated in the introduction to the IB Physics diploma programme, Physics seeks to explain nothing less than the Universe itself. In attempting to do this, the range of the magnitudes of various quantities will be huge.

If the numbers involved are going to mean anything, it is important to get some feel for their relative sizes. To avoid 'getting lost' amongst the numbers it is helpful to state them to the nearest **order of magnitude** or power of ten. The numbers are just rounded up or down as appropriate.

Comparisons can then be easily made because working out the ratio between two powers of ten is just a matter of adding or subtracting whole numbers. The diameter of an atom, 10^{-10} m, does not sound that much larger than the diameter of a proton in its nucleus, 10^{-15} m, but the ratio between them is 10^5 or 100000 times bigger. This is the same ratio as between the size of a railway station (order of magnitude 10^2 m) and the diameter of the Earth (order of magnitude 10^7 m).

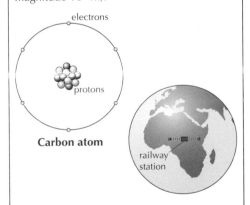

Carbon atom

Earth

For example, you would probably feel very pleased with yourself if you designed a new, environmentally friendly source of energy that could produce 2.03×10^3 J from 0.72 kg of natural produce. But the meaning of these numbers is not clear – is this a lot or is it a little? In terms of orders of magnitudes, this new source produces 10^3 joules per kilogram of produce. This does not compare terribly well with the 10^5 joules provided by a slice of bread or the 10^8 joules released per kilogram of petrol.

You do NOT need to memorise all of the values shown in the tables, but you should try and develop a familiarity with them. The ranges of magnitudes (first and last value) for the fundamental measurements of mass, length and time need to be known.

RANGE OF MASSES

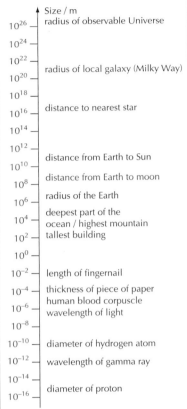

Mass / kg

10^{52}	total mass of observable Universe
10^{48}	
10^{44}	
10^{40}	mass of local galaxy (Milky Way)
10^{36}	mass of Sun
10^{32}	
10^{28}	
10^{24}	mass of Earth
10^{20}	total mass of oceans
10^{16}	total mass of atmosphere
10^{12}	
10^{8}	laden oil supertanker
10^{4}	elephant
10^{0}	human
	mouse
10^{-4}	
10^{-8}	grain of sand
10^{-12}	blood corpuscle
10^{-16}	bacterium
10^{-20}	
10^{-24}	haemoglobin molecule
10^{-28}	proton
10^{-32}	electron

RANGE OF LENGTHS

Size / m

10^{26}	radius of observable Universe
10^{24}	
10^{22}	
10^{20}	radius of local galaxy (Milky Way)
10^{18}	
10^{16}	distance to nearest star
10^{14}	
10^{12}	
10^{10}	distance from Earth to Sun
10^{8}	distance from Earth to moon
10^{6}	radius of the Earth
10^{4}	deepest part of the ocean / highest mountain
10^{2}	tallest building
10^{0}	
10^{-2}	length of fingernail
10^{-4}	thickness of piece of paper
	human blood corpuscle
10^{-6}	wavelength of light
10^{-8}	
10^{-10}	diameter of hydrogen atom
10^{-12}	wavelength of gamma ray
10^{-14}	
10^{-16}	diameter of proton

RANGE OF TIMES

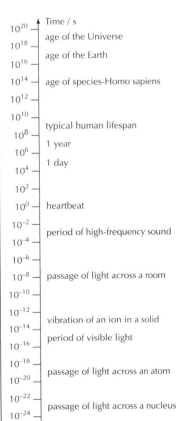

Time / s

10^{20}	age of the Universe
10^{18}	age of the Earth
10^{16}	
10^{14}	age of species-Homo sapiens
10^{12}	
10^{10}	
10^{8}	typical human lifespan
10^{6}	1 year
10^{4}	1 day
10^{2}	
10^{0}	heartbeat
10^{-2}	
10^{-4}	period of high-frequency sound
10^{-6}	
10^{-8}	passage of light across a room
10^{-10}	
10^{-12}	
10^{-14}	vibration of an ion in a solid
10^{-16}	period of visible light
10^{-18}	
10^{-20}	passage of light across an atom
10^{-22}	
10^{-24}	passage of light across a nucleus

RANGE OF ENERGIES

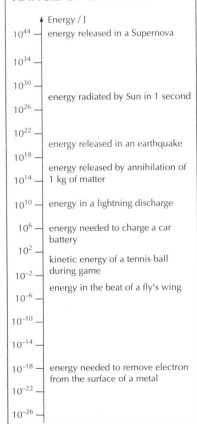

Energy / J

10^{44}	energy released in a Supernova
10^{34}	
10^{30}	energy radiated by Sun in 1 second
10^{26}	
10^{22}	energy released in an earthquake
10^{18}	energy released by annihilation of 1 kg of matter
10^{14}	
10^{10}	energy in a lightning discharge
10^{6}	energy needed to charge a car battery
10^{2}	kinetic energy of a tennis ball during game
10^{-2}	energy in the beat of a fly's wing
10^{-6}	
10^{-10}	
10^{-14}	
10^{-18}	energy needed to remove electron from the surface of a metal
10^{-22}	
10^{-26}	

The SI system of fundamental and derived units

FUNDAMENTAL UNITS

Any measurement and every quantity can be thought of as being made up of two important parts.

1. The number and
2. The units.

Without **both** parts, the measurement does not make sense. For example a person's age might be quoted as 'seventeen years old' but without the 'years' the situation is not clear. Are they 17 minutes, 17 months or 17 years old? In this case you would know if you saw them, but a statement like

length = 4.2

actually says nothing. Having said this, it is really surprising to see the number of candidates who forget to include the units in their answers to examination questions.

In order for the units to be understood, they need to be defined. There are many possible systems of measurement that have been developed. In science we use the International System of units (SI). In SI, the **fundamental** or **base** units are as follows

Quantity	SI unit	SI symbol
Mass	kilogram	kg
Length	metre	m
Time	second	s
Electric current	ampere	A
Amount of substance	mole	mol
Temperature	kelvin	K
(Luminous intensity	candela	cd)

You do not need to know the precise definitions of any of these units in order to use them properly.

DERIVED UNITS

Having fixed the fundamental units, all other measurements can be expressed as different combinations of the fundamental units. In other words, all the other units are **derived units**. For example, the fundamental list of units does not contain a unit for the measurement of speed. The definition of speed can be used to work out the derived unit.

$$\text{Since speed} = \frac{\text{distance}}{\text{time}}$$

$$\text{Units of speed} = \frac{\text{units of distance}}{\text{units of time}}$$

$$= \frac{\text{metres}}{\text{seconds}} \text{ (pronounced 'metres per second')}$$

$$= \frac{m}{s}$$

$$= m\,s^{-1}$$

Of the many ways of writing this unit, the last way ($m\,s^{-1}$) is the best.

Sometimes particular combinations of fundamental units are so common that they are given a new derived name. For example, the unit of force is a derived unit – it turns out to be $kg\,m\,s^{-2}$. This unit is given a new name the newton (N) so that $1N = 1\ kg\,m\,s^{-2}$.

The great thing about SI is that, so long as the numbers that are substituted into an equation are in SI units, then the answer will also come out in SI units. You can always 'play safe' by converting all the numbers into proper SI units. Sometimes, however, this would be a waste of time.

There are some situations where the use of SI becomes awkward. In astronomy, for example, the distances involved are so large that the SI unit (the metre) always involves large orders of magnitudes. In these cases, the use of a different (but non SI) unit is very common. Astronomers can use the astronomical unit (AU), the light-year (ly) or the parsec (pc) as appropriate. Whatever the unit, the conversion to SI units is simple arithmetic.

$$1\ AU = 1.5 \times 10^{11}\ m$$

$$1\ ly = 9.5 \times 10^{15}\ m$$

$$1\ pc = 3.1 \times 10^{16}\ m$$

There are also some units (for example the hour) which are so common that they are often used even though they do not form part of SI. Once again, before these numbers are substituted into equations they need to be converted.

The table below lists the SI derived units that you will meet.

SI derived unit	SI base unit	Alternative SI unit
newton (N)	$kg\,m\,s^{-2}$	–
pascal (Pa)	$kg\,m^{-1}\,s^{-2}$	$N\,m^{-2}$
hertz (Hz)	s^{-1}	–
joule (J)	$kg\,m^2\,s^{-2}$	$N\,m$
watt (W)	$kg\,m^2\,s^{-3}$	$J\,s^{-1}$
coulomb (C)	$A\,s$	–
volt (V)	$kg\,m^2\,s^{-3}\,A^{-1}$	W/A
ohm (Ω)	$kg\,m^2\,s^{-3}\,A^{-2}$	V/A
weber (Wb)	$kg\,m^2\,s^{-2}\,A^{-1}$	$V\,s$
tesla (T)	$kg\,s^{-2}\,A^{-1}$	$Wb\,m^{-2}$
becquerel (Bq)	s^{-1}	–
gray (Gy)	$m^2\,s^{-2}$	$J\,kg^{-1}$
sievert (Sv)	$m^2\,s^{-2}$	$J\,kg^{-1}$

PREFIXES

To avoid the repeated use of standard form, an alternative is to use one of the list of agreed prefixes given in the IB data booklet. These can be very useful but they can also lead to errors in calculations. It is very easy to forget to include the conversion factor.

For example, 1 kW = 1000 W. 1 mW = 10^{-3} W (in other words, $\frac{1W}{1000}$)

Uncertainties and error in experimental measurement

ERRORS – RANDOM AND SYSTEMATIC (PRECISION AND ACCURACY)

An experimental error just means that there is a difference between the recorded value and the 'perfect' or 'correct' value. Errors can be categorised as **random** or **systematic**.

Repeating readings does not reduce systematic errors.

Sources of random errors include

- The readability of the instrument
- The observer being less than perfect
- The effects of a change in the surroundings

Sources of systematic errors include

- An instrument with **zero error**. To correct for zero error the value should be subtracted from every reading.
- An instrument being wrongly **calibrated**
- The observer being less than perfect in the same way every measurement

An **accurate** experiment is one that has a small systematic error, whereas a **precise** experiment is one that has a small random error.

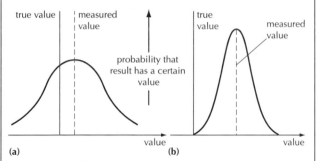

(a)   **(b)**

Two examples illustrating the nature of experimental results:
(a) an accurate experiment of low precision
(b) a less accurate but more precise experiment.

GRAPHICAL REPRESENTATION OF UNCERTAINTY

In many situations the best method of presenting and analysing data is to use a graph. If this is the case, a neat way of representing the uncertainties is to use **error bars**. The graphs below explains their use.

Since the error bar represents the uncertainty range, the 'best-fit' line of the graph should pass through ALL of the rectangles created by the error bars.

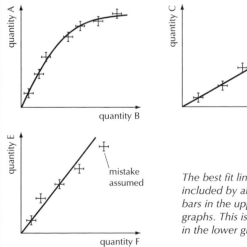

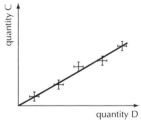

The best fit line is included by all the error bars in the upper two graphs. This is not true in the lower graph.

Systematic and random errors can often be recognised from a graph of the results.

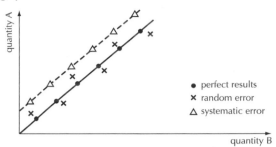

- ● perfect results
- ✕ random error
- △ systematic error

Perfect results, random and systematic errors of two proportional quantities.

ESTIMATING THE UNCERTAINTY RANGE

An **uncertainty range** applies to any experimental value. The idea is that, instead of just giving one value that implies perfection, we give the likely range for the measurement.

1. Estimating from first principles
All measurement involves a readability error. If we use a measuring cylinder to find the volume of a liquid, we might think that the best estimate is 73 cm³, but we know that it is not exactly this value (73.00000000000 cm³).

Uncertainty range is ± 5 cm³.
We say volume = 73 ± 5 cm³.

Normally the uncertainty range due to readability is estimated as below.

Device	Example	Uncertainty
Analogue scale	Rulers, meters with moving pointers	± (half the smallest scale division)
Digital scale	Top-pan balances, digital meters	± (the smallest scale division)

2. Estimating uncertainty range from several repeated measurements
If the time taken for a trolley to go down a slope is measured five times, the readings in seconds might be 2.01, 1.82, 1.97, 2.16 and 1.94. The average of these five readings is 1.98 s. The deviation of the largest and smallest readings can be calculated (2.16 − 1.98 = 0.18; 1.98 − 1.82 = 0.16). The largest value is taken as the uncertainty range. In this example the time is 1.98 s ± 0.18 s

SIGNIFICANT DIGITS

Any experimental measurement should be quoted with its uncertainty. This indicates the possible range of values for the quantity being measured. At the same time, the number of **significant digits** used will act as a guide to the amount of uncertainty. For example, a measurement of mass which is quoted as 23.456 g implies an uncertainty of ± 0.001 g (it has five significant digits), whereas one of 23.5 g implies an uncertainty of ± 0.1 g (it has three significant digits).

A simple rule for calculations (multiplication or division) is to quote the answer to the same number of significant digits as the LEAST precise value that is used.

For a more complete analysis of how to deal with uncertainties in calculated results, see page 57.

Estimation

ORDERS OF MAGNITUDE

It is important to develop a 'feeling' for some of the numbers that you use. When using a calculator, it is very easy to make a simple mistake (e.g. by entering the data incorrectly). A good way of checking the answer is to first make an estimate before resorting to the calculator. The multiple-choice paper (paper 1) does not allow the use of calculators.

Approximate values for each of the fundamental SI units are given below.

1 kg	A packet of sugar, 1 litre of water. A person would be about 50 kg or more.
1 m	Distance between one's hands with arms outstretched
1 s	Duration of a heart beat (when resting – it can easily double with exercise)
1 amp	Current flowing from the mains electricity when a computer is connected. The maximum current to a domestic device would be about 10 A or so.
1 kelvin	1K is a very low temperature. Water freezes at 273 K and boils at 373 K. Room temperature is about 300 K.
1 mol	12 g of carbon–12. About the number of atoms of carbon in the 'lead' of a pencil.

The same process can happen with some of the derived units.

1 m s^{-1}	Walking speed. A car moving at 30 m s^{-1} would be fast
1 m s^{-2}	Quite a slow acceleration. The acceleration of gravity is 10 m s^{-2}
1 N	A small force – about the weight of an apple
1 V	Batteries generally range from a few volts up to 20 or so, the mains is several hundred volts
1 Pa	A very small pressure. Atmospheric pressure is about 10^5 Pa.
1 J	A very small amount of energy – the work done lifting a apple off the ground

POSSIBLE REASONABLE ASSUMPTIONS

Everyday situations are very complex. In Physics we often simplify a problem by making simple assumptions. Even if we know these assumptions are not absolutely true they allow us to gain an understanding of what is going on. At the end of the calculation it is often possible to go back and work out what would happen if our assumption turned out not to be true.

The table below lists some common assumptions. Be careful not to assume too much! Additionally we often have to assume that some quantity is constant even if we know that in reality it is varying slightly all the time.

Assumption	Example
Friction is negligible	Many mechanics situations – but need to be very careful.
No heat lost	Almost all thermal situations
Mass of connecting string etc is negligible	Many mechanics situations
Resistance of ammeter is zero	Circuits
Resistance of voltmeter is infinite	Circuits
Internal resistance of battery is zero	Circuits
Material obeys Ohm's law	Circuits
Machine 100% efficient	Many situations
Gas is ideal	Thermodynamics
Collision is elastic	Only gas molecules have perfectly elastic collisions

CALCULUS NOTATION

If one wants to mathematically analyse motion in detail, the correct way to do this would be to use a branch of mathematics called calculus. A knowledge of the details of calculus is **not** required for the IB physics course. However sometimes it does help to be able to use calculus shorthand.

Symbol	Pronounced	Meaning	Example
Δx	'Delta x'	= 'the change in x'	Δt means 'the change in time'
δx	'Delta x'	= 'the small change in x'	δt means 'the small change in time'
$\dfrac{\Delta x}{\Delta t}$	'Delta x divided by delta t' or 'Delta x over delta t'	= 'the AVERAGE rate of change of x' This average is calculated over a relatively large period of time	$\dfrac{\Delta s}{\Delta t}$ often means the average speed
$\dfrac{\delta x}{\delta t}$	'Delta x divided by delta t' or 'Delta x over delta t'	= 'the AVERAGE rate of change of x' This average is calculated over a small period of time	$\dfrac{\delta s}{\delta t}$ means the average speed
$\dfrac{dx}{dt}$	'Dee x by dee t' or Dee by dee t of x	= 'the INSTANTANEOUS rate of change of x' This value is calculated at one instant of time	$\dfrac{ds}{dt}$ means the instantaneous speed

Graphs

PLOTTING GRAPHS – AXES AND BEST FIT

The reason for plotting a graph in the first place is that it allows us to identify trends. To be precise, it allows us a visual way of representing the variation of one quantity with respect to another. When plotting graphs, you need to make sure that all of the following points have been remembered

- The graph should have a title. Sometimes they also need a key.
- The scales of the axes should be suitable – there should not, of course, be any sudden or uneven 'jumps' in the numbers.
- The inclusion of the origin has been thought about. Most graphs should have the origin included – it is rare for a graph to be improved by this being missed out. If in doubt include it. You can always draw a second graph without it if necessary.
- The final graph should, if possible, cover more than half the paper in either direction.
- The axes are labelled with both the quantity (e.g. current) AND the units (e.g.amps).

- The points are clear. Vertical and horizontal lines to make crosses are better than 45 degree crosses or dots.
- All the points have been plotted correctly.
- Error bars are included if appropriate.
- A best-fit trend line is added. This line NEVER just 'joins the dots' – it is there to show the overall trend.
- If the best-fit line is a curve, this has been drawn as a single smooth line.
- If the best-fit line is a straight line, this has been added WITH A RULER
- As a general rule, there should be roughly the same number of points above the line as below the line.
- Check that the points are randomly above and below the line. Sometimes people try to fit a best-fit straight line to points that should be represented by a gentle curve. If this was done then points below the line would be at the beginning of the curve and all the points above the line would be at the end, or vice versa.
- Any points that do not agree with the best-fit line have been identified.

MEASURING INTERCEPT, GRADIENT AND AREA UNDER.

Graphs can be used to analyse the data. This is particularly easy for straight-line graphs, though many of the same principles can be used for curves as well. Three things are particularly useful: the **intercept**, the **gradient** and the **area under the graph**.

1. Intercept

In general, a graph can intercept (cut) either axis any number of times. A straight-line graph can only cut each axis once and often it is the **y-intercept** that has particular importance. (Sometimes the y-intercept is referred to as simply 'the intercept'.) If a graph has an intercept of zero it goes through the origin. **Proportional** – note that two quantities are proportional if the graph is a straight line THAT PASSES THROUGH THE ORIGIN.

Sometimes a graph has to be 'continued on' (outside the range of the readings) in order for the intercept to be found. This process is known as **extrapolation**. The process of assuming that the trend line applies between two points is known as **interpolation**.

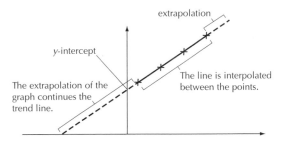

2. Gradient

The gradient of a straight-line graph is the increase in the y-axis value divided by the increase in the x-axis value.

The following points should be remembered

- A straight-line graph has a constant gradient.
- The triangle used to calculate the gradient should be as large as possible.
- The gradient has units. It is the units on the y-axis divided by the units on the x-axis.

- Only if the x-axis is a measurement of time does the gradient represent the RATE at which the quantity on the y-axis increases.

The gradient of a curve at any particular point is the gradient of the tangent to the curve at that point.

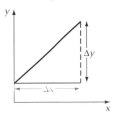 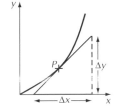

gradient of straight line $= \frac{\Delta y}{\Delta x}$ at point P on the curve, gradient $= \frac{\Delta y}{\Delta x}$

3. Area under a graph

The area under a straight-line graph is the product of multiplying the average quantity on the y-axis by the quantity on the x-axis. This does not always represent a useful physical quantity. When working out the area under the graph

- If the graph consists of straight-line sections, the area can be worked out by dividing the shape up into simple shapes.
- If the graph is a curve, the area can be calculated by 'counting the squares' and working out what one square represents.
- The units for the area under the graph are the units on the y-axis multiplied by the units on the x-axis.
- If the mathematical equation of the line is known, the area of the graph can be calculated using a process called **integration**.

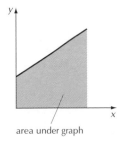 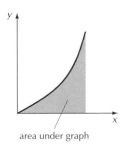

area under graph   area under graph

Graphical analysis and determination of relationships

EQUATION OF A STRAIGHT-LINE GRAPH

All straight-line graphs can be described using one general equation

$$y = mx + c$$

y and x are the two variables (to match with the y-axis and the x-axis).
m and c are both constants – they have one fixed value.

- c represents the intercept on the y-axis (the value y takes when $x = 0$)
- m is the gradient of the graph.

In some situations, a direct plot of the measured variable will give a straight line. In some other situations we have to choose carefully what to plot in order to get a straight line. In either case, once we have a straight line, we then use the gradient and the intercept to calculate other values.

For example, a simple experiment might measure the velocity of a trolley as it rolls down a slope. The equation that describes the motion is $v = u + at$ where u is the initial velocity of the object. In this situation v and t are our variables, a and u are the constants.

You should be able to see that the physics equation has exactly the same form as the mathematical equation. The order has been changed below so as to emphasis the link.

$$v = u + at$$
$$y = c + mx$$

By comparing these two equations, you should be able to see that if we plot the velocity on the y-axis and the time on the x-axis we are going to get a straight-line graph.

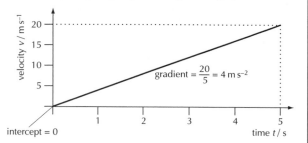

The comparison also works for the constants.

- c (the y-intercept) must be equal to the initial velocity u
- m (the gradient) must be equal to the acceleration a

In this example the graph tells us that the trolley must have started from rest (intercept zero) and it had a constant acceleration of 4.0 m s^{-2}.

CHOOSING WHAT TO PLOT TO GET A STRAIGHT LINE

With a little rearrangement we can often end up with the physics equation in the same form as the mathematical equation of a straight line. Important points include

- Identify which symbols represent variables and which symbols represent constants.
- The symbols that correspond to x and y must be variables and the symbols that correspond to m and c must be constants.
- If you take a variable reading and square it (cube, square root, reciprocal etc) – the result is still a variable and you could choose to plot this on one of the axes.
- You can plot any mathematical combination of your original readings on one axis – this is still a variable.
- Sometimes the physical quantities involved use the symbols m (e.g. mass) or c (e.g. speed of light). Be careful not to confuse these with the symbols for gradient or intercept.

Example 1

The gravitational force F that acts on an object at a distance r away from the centre of a planet is given by the equation

$$F = \frac{GMm}{r^2}$$ where M is the mass of the planet and the m is mass of the object.

If we plot force against distance we get a curve (graph 1).

We can restate the equation as $F = \frac{GMm}{r^2} + 0$

and if we plot F on the y-axis and $\frac{1}{r^2}$ on the x-axis we will get a straight-line (graph 2).

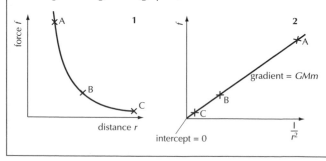

Example 2

If an object is placed in front of a lens we get an image. The image distance v is related to the object distance u and the focal length of the lens f by the following equation.

$$\frac{1}{u} + \frac{1}{v} = \frac{1}{f}$$

There are many possible ways to rearrange this in order to get it into straight–line form. You should check that all these are algebraically the same.

$$v + u = \frac{uv}{f} \quad \text{or} \quad \frac{v}{u} = \frac{v}{f} - 1 \quad \text{or} \quad \frac{1}{u} = \frac{1}{f} - \frac{1}{v}$$

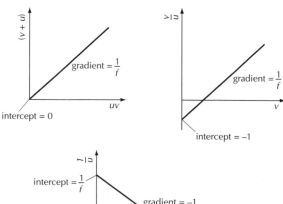

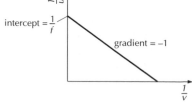

Vectors and Scalars

DIFFERENCE BETWEEN VECTORS AND SCALARS

If you measure any quantity, it must have a number AND a unit. Together they express the **magnitude** of the quantity. Some quantities also have a direction associated with them. A quantity that has magnitude and direction is called a **vector** quantity whereas one that has only magnitude is called a **scalar** quantity. For example, all forces are vectors.

The table lists some common quantities. The first two quantities in the table are linked to one another by their definitions (see page 9). All the others are in no particular order.

Vectors	Scalars
Displacement ⟷	Distance
Velocity ⟷	Speed
Acceleration	Mass
Force	Energy (all forms)
Momentum	Temperature
Electric field strength	Potential or potential difference
Magnetic field strength	Density
Gravitational field strength	Area

Although the vectors used in many of the given examples are forces, the techniques can be applied to all vectors.

COMPONENTS OF VECTORS

It is also possible to 'split' one vector into two (or more) vectors. This process is called **resolving** and the vectors that we get are called the **components** of the original vector. This can be a very useful way of analysing a situation if we choose to resolve all the vectors into two directions that are at right angles to one another.

 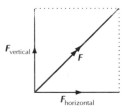

Splitting a vector into components

These 'mutually perpendicular' directions are totally independent of each other and can be analysed separately. If appropriate, both directions can then be combined at the end to work out the final resultant vector.

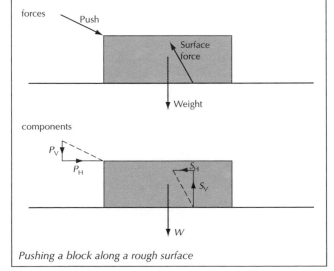

Pushing a block along a rough surface

REPRESENTING VECTORS

In most books (including the data booklet that you are allowed to use in the examination) a bold letter is used to represent a vector whereas a normal letter represents a scalar. For example F would be used to represent a force in magnitude AND direction. The list below shows some other recognised methods.

$$\vec{F}, \ \bar{F} \ \text{or} \ \underline{F}$$

Vectors are best shown in diagrams using arrows:

- the relative magnitudes of the vectors involved are shown by the relative length of the arrows
- the direction of the vectors is shown by the direction of the arrows.

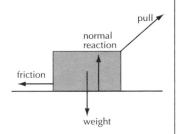

ADDITION / SUBTRACTION OF VECTORS

If we have a 3N and a 4N force, the overall force (resultant force) can be anything between 1N and 7N depending on the directions involved.

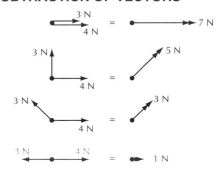

The way to take the directions into account is to do a scale diagram and use the parallelogram law of vectors.

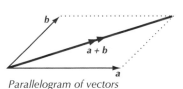

Parallelogram of vectors

This process is the same as adding vectors in turn – the 'tail' of one vector is drawn starting from the head of the previous vector.

TRIGONOMETRY

Vector problems can always been solved using scale diagrams, but this can be very time consuming. The mathematics of trigonometry often makes it much easier to use the mathematical functions of sine or cosine. This is particularly appropriate when resolving. The diagram below shows how to calculate the values of either of these components.

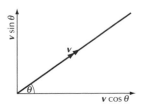

See page 12 for an example.

IB QUESTIONS – PHYSICS AND PHYSICAL MEASUREMENT

1 Which one of the following quantities is a scalar?

   A Weight

   B Distance

   C Velocity

   D Momentum

2 Which one of the following is a fundamental unit in the International System of units (S.I.)?

   A newton

   B ampere

   C joule

   D pascal

3 Gravitational field strength may be specified in N kg^{-1}. Units of N kg^{-1} are equivalent to

   A m s^{-1}

   B m s^{-2}

   C kg m s^{-1}

   D kg m s^{-2}

4 Which one of the following is a scalar quantity?

   A Electric field

   B Acceleration

   C Power

   D Momentum

5 A motor car travels on a circular track of radius, a, as shown in the figure. When the car has travelled from P to Q its displacement from P is

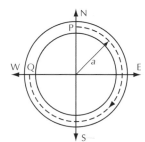

   A $a\sqrt{2}$ southwest

   B $a\sqrt{2}$ northeast

   C $\frac{3\pi a}{2}$ southwest

   D $\frac{3\pi a}{2}$ northeast

6 The frequency of oscillation f of a mass m suspended from a vertical spring is given by

$$f = \frac{1}{2\pi}\sqrt{\frac{k}{m}}$$

where k is the spring constant.

Which **one** of the following plots would produce a straight-line graph?

   A f against m

   B f^2 against $\frac{1}{m}$

   C f against $\sqrt{m}$

   D $\frac{1}{f}$ against m

7 Repeated measurements of a quantity can reduce the effects of

   A both random and systematic errors

   B only random errors

   C only systematic errors

   D neither random nor systematic errors

8 A vector v makes an angle θ with the x axis as shown.

As the angle θ increases from $0°$ to $90°$, how do the x and y components of v vary?

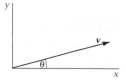

	x component	y component
A	Increases	Increases
B	Increases	Decreases
C	Decreases	Increases
D	Decreases	Decreases

9 An object is rolled from rest down an inclined plane. The distance travelled by the object was measured at seven different times. A graph was then constructed of the distance travelled against the (time taken)^2 as shown below.

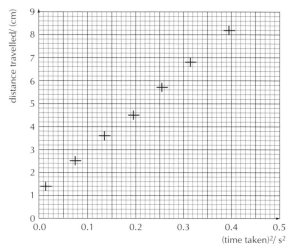

(a) **(i)** What quantity is given by the gradient of such a graph? [2]

     (ii) Explain why the graph suggests that the collected data is valid but includes a **systematic error**. [2]

     (iii) Do these results suggest that distance is proportional to (time taken)^2? Explain your answer. [2]

     (iv) Making allowance for the systematic error, calculate the acceleration of the object. [2]

(b) The following graph shows that same data after the uncertainty ranges have been calculated and drawn as error bars.

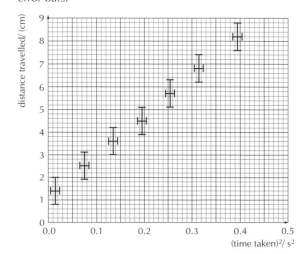

Add two lines to show the range of the possible acceptable values for the gradient of the graph. [2]

Kinematic concepts

DEFINITIONS

These technical terms should not be confused with their 'everyday' use. In particular one should note that

- vector quantities always have a direction associated with them.
- generally, velocity and speed are NOT the same thing. This is particularly important if the object is not going in a straight line.
- the units of acceleration come from its definition. $(\text{m s}^{-1}) \div \text{s} = \text{m s}^{-2}$.
- the definition of acceleration is precise. It is related to the change in **velocity** (not the same thing as the change in speed). Whenever the motion of an object changes, it is called acceleration. For this reason acceleration does not necessarily mean constantly increasing speed – it is possible to accelerate while at constant speed if the direction is changed.
- a deceleration is just a negative acceleration.

	Symbol	Definition	Example	SI Unit	Vector or scalar?
Displacement	s	The distance moved in a particular direction	The displacement from London to Rome is 1.43×10^6 m southeast.	m	Vector
Velocity	v or u	The rate of change of displacement. $\text{velocity} = \dfrac{\text{change of displacement}}{\text{time taken}}$	The average velocity during a flight from London to Rome is 160 m s^{-1} southeast.	m s^{-1}	Vector
Speed	v or u	The rate of change of distance. $\text{speed} = \dfrac{\text{distance gone}}{\text{time taken}}$	The average speed during a flight from London to Rome is 160 m s^{-1}	m s^{-1}	Scalar
Acceleration	a	The rate of change of velocity $\text{acceleration} = \dfrac{\text{change of velocity}}{\text{time taken}}$	The average acceleration of a plane on the runway during take-off is 3.5 m s^{-2} in a forwards direction. This means that on average, its velocity changes every second by 3.5 m s^{-1}	m s^{-2}	Vector

INSTANTANEOUS VS AVERAGE

It should be noticed that the **average** value (over a period of time) is very different to the **instantaneous** value (at one particular time).

In the example below, the positions of a sprinter are shown at different times after the start of a race.

The average speed over the whole race is easy to work out. It is the total distance (100 m) divided by the total time (11.3 s) giving 8.5 m s^{-1}.

But during the race, her instantaneous speed was changing all the time. At the end of the first 2.0 seconds, she had travelled 10.04 m. This means that her average speed over the first 2.0 seconds was 5.02 m s^{-1}. During these first two seconds, her instantaneous speed was increasing – she was accelerating. If she started at rest (speed = 0.00 m s^{-1}) and her **average** speed (over the whole two seconds) was 5.02 m s^{-1} then her instantaneous speed at 2 seconds must be more than this. In fact the instantaneous speed for this sprinter was 9.23 m s^{-1}, but it would not be possible to work this out from the information given.

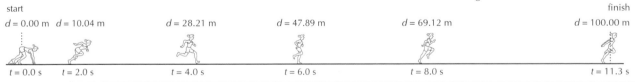

start   finish

$d = 0.00$ m $d = 10.04$ m $d = 28.21$ m $d = 47.89$ m $d = 69.12$ m $d = 100.00$ m

$t = 0.0$ s $t = 2.0$ s $t = 4.0$ s $t = 6.0$ s $t = 8.0$ s $t = 11.3$ s

FRAMES OF REFERENCE

If two things are moving in the same straight line but are travelling at different speeds, then we can work out their **relative velocities** by simple addition or subtraction as appropriate. For example, imagine two cars travelling along a straight road at different speeds.

If one car (travelling at 30 m s^{-1}) overtakes the other car (travelling at 25 m s^{-1}), then according to the driver of the slow car, the relative velocity of the fast car is + 5 m s^{-1}.

In technical terms what we are doing is moving from one **frame of reference** into another. The velocities of 25 m s^{-1} and 30 m s^{-1} were measured according

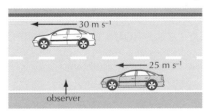

Above: one car overtaking another, as seen by an observer on the side of the road. Right: one car overtaking another, as seen by the driver of the slow car.

to a stationary observer on the side of the road. We moved from this frame of reference into the driver's frame of reference.

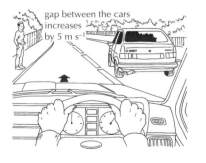

Graphical representation of motion

THE USE OF GRAPHS

Graphs are very useful for representing the changes that happen when an object is in motion. There are three possible graphs that can provide useful information

- displacement – time or distance – time graphs
- velocity – time or speed – time graphs
- acceleration – time graphs.

There are two common methods of determining particular physical quantities from these graphs. The particular physical quantity determined depends on what is being plotted on the graph.

1. Finding the gradient of the line.

To be a little more precise, one could find either the gradient of
- a straight-line section of the graph (this finds an average value), or
- the tangent to the graph at one point (this finds an instantaneous value).

2. Finding the area under the line.

To make things simple at the beginning, the graphs are normally introduced by considering objects that are just moving in one particular direction. If this is the case then there is not much difference between the scalar versions (distance or speed) and the vector versions (displacement or velocity) as the directions are clear from the situation. More complicated graphs can look at the component of a velocity in a particular direction.

If the object moves forward then backward (or up then down), we distinguish the two velocities by choosing which direction to call positive. It does not matter which direction we choose, but it should be clearly labelled on the graph.

Many examination candidates get the three types of graph muddled up. For example a speed – time graph might be interpreted as a distance – time graph or even an acceleration – time graph. Always look at the axes of a graph very carefully.

DISPLACEMENT – TIME GRAPHS

- The gradient of a displacement – time graph is the velocity
- The area under a displacement – time graph does not represent anything useful

Examples

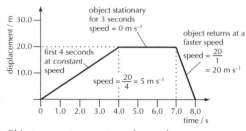

Object moves at constant speed, stops then returns.

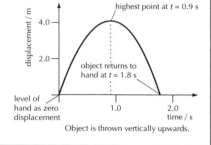

Object is thrown vertically upwards.

VELOCITY – TIME GRAPHS

- The gradient of a velocity – time graph is the acceleration
- The area under a velocity – time graph is the displacement

Examples

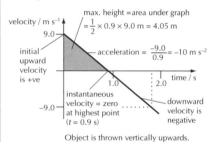

Object is thrown vertically upwards.

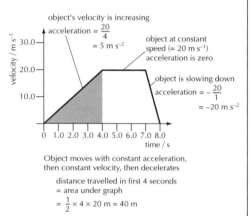

Object moves with constant acceleration, then constant velocity, then decelerates

distance travelled in first 4 seconds
= area under graph
= $\frac{1}{2} \times 4 \times 20$ m = 40 m

ACCELERATION-TIME GRAPHS

- The gradient of an acceleration – time graph is not often useful (it is actually the rate of change of acceleration).
- The area under an acceleration – time graph is the change in velocity

Examples

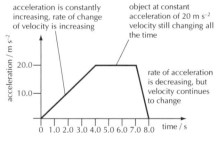

Object moves with increasing, then constant, then decreasing acceleration.

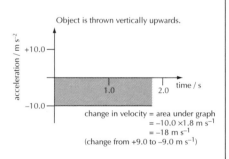

change in velocity = area under graph
= -10.0×1.8 m s^{-1}
= -18 m s^{-1}
(change from +9.0 to –9.0 m s^{-1})

EXAMPLE OF EQUATION OF UNIFORM MOTION

A car accelerates uniforms from rest. After 8 s it has travelled 120 m.
Calculate: (i) its average acceleration   (ii) its instantaneous speed after 8 s

(i) $s = ut + \frac{1}{2} at^2$

∴ $120 = 0 \times 8 + \frac{1}{2} a \times 8^2 = 32 a$

$a = 3.75$ m s^{-2}

(ii) $v^2 = u^2 + 2 as$

$= 0 + 2 \times 3.75 \times 120$

$= 900$

∴ $v = 30$ m s^{-1}

Uniformly accelerated motion

PRACTICAL CALCULATIONS

In order to determine how the velocity (or the acceleration) of an object varies in real situations, it is often necessary to record its motion. Possible laboratory methods include

Light gates

A light gate is a device that senses when an object cuts through a beam of light. The time for which the beam is broken is recorded. If the length of the object that breaks the beam is known, the average speed of the object through the gate can be calculated.

Alternatively, two light gates and a timer can be used to calculate the average velocity between the two gates. Several light gates and a computer can be joined together to make direct calculations of velocity or acceleration.

Strobe photography

A strobe light gives out very brief flashes of light at fixed time intervals. If a camera is pointed at an object and the only source of light is the strobe light, then the developed picture will have captured an object's motion.

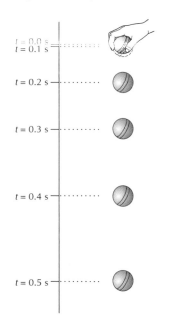

Ticker timer

A ticker timer can be arranged to make dots on a strip of paper at regular intervals of time (typically every fiftieth of a second). If the piece of paper is attached to an object, and the object is allowed to fall, the dots on the strip will have recorded the distance moved by the object in a known time.

EQUATIONS OF UNIFORM MOTION

These equations can only be used when the acceleration is constant – don't forget to check if this is the case!

The list of variables to be considered (and their symbols) is as follows

- u initial velocity
- v final velocity
- a acceleration (constant)
- t time taken
- s distance travelled

The following equations link these different quantities.

$$v = u + at$$
$$s = \left(\frac{u + v}{2}\right) t$$
$$v^2 = u^2 + 2as$$
$$s = ut + \frac{1}{2}at^2$$
$$s = vt - \frac{1}{2}at^2$$

The first equation is derived from the definition of acceleration. In terms of these symbols, this definition would be

$$a = \frac{(v - u)}{t}$$

This can be rearranged to give the first equation.

$$v = u + at \quad (1)$$

The second equation comes from the definition of average velocity.

$$\text{average velocity} = \frac{s}{t}$$

Since the velocity is changing uniformly we know that this average velocity must be given by

$$\text{average velocity} = \frac{(v + u)}{2}$$

or $\quad \frac{s}{t} = \frac{(u + v)}{2}$

This can be rearranged to give

$$s = \frac{(u + v)t}{2} \quad (2)$$

The other equations of motion can be derived by using these two equations and substituting for one of the variables (see previous page for example).

FALLING OBJECTS

A very important example of uniformly accelerated motion is the vertical motion of an object in a uniform **gravitational field**. If we ignore the effects of air resistance, this is known as being in **free-fall**.

Taking down as positive, the graphs of the motion of any object in free-fall are

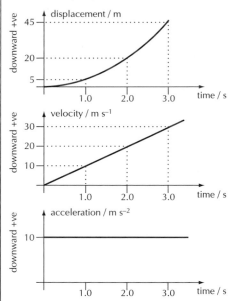

In the absence of air resistance, all falling objects have the SAME acceleration of free-fall, INDEPENDENT of their mass.

Air resistance will (eventually) affect the motion of all objects. Typically, the graphs of a falling object affected by air resistance become the shapes shown below

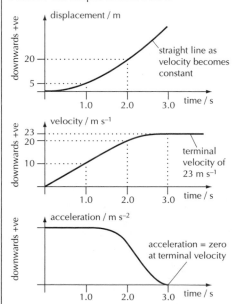

As the graphs show, the velocity does not keep on rising. It eventually reaches a maximum or **terminal velocity**. A piece of falling paper will reach its terminal velocity in a much shorter time than a falling book.

Forces and free-body diagrams

FORCES – WHAT THEY ARE AND WHAT THEY DO

In the examples below, a force (the kick) can cause deformation (the ball changes shape) or a change in motion (the ball gains a velocity). There are many different types of forces, but in general terms one can describe any force as 'the cause of a deformation or a velocity change'. The SI unit for the measurement of forces is the newton (N).

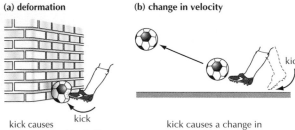

(a) deformation

kick causes
deformation of football

kick

(b) change in velocity

kick

kick causes a change in
motion of football

Effect of a force on a football

- A (resultant) force causes a CHANGE in velocity. If the (resultant) force is zero then the velocity is constant. Remember a change in velocity is called an acceleration, so we can say that **a force causes an acceleration**. A (resultant) force is NOT needed for a constant velocity (see page 13).
- The fact that a force can cause deformation is also important, but the deformation of the ball was, in fact, not caused by just one force – there was another one from the wall.
- One force can act on only one object. To be absolutely precise the description of a force should include
  - its magnitude
  - its direction
  - the object on which it acts (or the part of a large object)
  - the object that exerts the force
  - the nature of the force

A description of the force shown in the example would thus be 'a 50 N push at 20° to the horizontal acting ON the football FROM the boot'.

DIFFERENT TYPES OF FORCES

The following words all describe the forces (the pushes or pulls) that exist in nature

Gravitational force	**Normal reaction**	**Compression**
Electrostatic force	**Friction**	**Upthrust**
Magnetic force	**Tension**	**Lift**

One way of categorising these forces is whether they result from the contact between two surfaces or whether the force exists even if a distance separates the objects.

The origin of all these everyday forces is either gravitational or electromagnetic. The vast majority of everyday effects that we observe are due to electromagnetic forces.

MEASURING FORCES

The simplest experimental method for measuring the size of a force is to use the **extension** of a spring. When a spring is in tension it increases in length. The difference between the natural length and stretched length is called the extension of a spring.

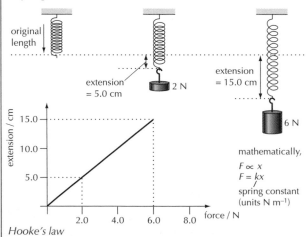

original
length

extension
= 5.0 cm

2 N

extension
= 15.0 cm

6 N

mathematically,
$F \propto x$
$F = kx$
spring constant
(units N m^{-1})

Hooke's law

Hooke's law states that up to the elastic limit, the extension, x, of a spring is proportional to the tension force, F. The constant of proportionality k is called the **spring constant**. The SI units for the spring constant are N m^{-1}. Thus by measuring the extension, we can calculate the force.

FORCES AS VECTORS

Since forces are vectors, vector mathematics must be used to find the resultant force from two or more other forces. A force can also be split into its components. See page 7 for more details.

(a) by vector mathematics
example: block being pushed on rough surface

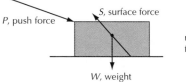

S, surface force

P, push force

W, weight

force diagram:

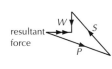

W

S

resultant
force

P

(b) by components
example: block sliding down a smooth slope

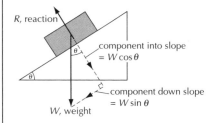

R, reaction

component into slope
$= W \cos \theta$

component down slope
$= W \sin \theta$

W, weight

resultant down
slope $= W \sin \theta$

resultant into
slope $= W \cos \theta - R$
$= $ zero

Vector addition

FREE-BODY DIAGRAMS

In a **free-body diagram**,
- one object (and ONLY one object) is chosen.
- all the forces on that object are shown and labelled

For example, if we considered the simple situation of a book resting on a table, we can construct free-body diagrams for either the book or the table.

situation:

free-body diagram for book:

R_T, reaction from table

w, weight of book
gravitational pull of Earth

free-body diagram for table:

R_E

P, push from
book

R_E, reaction
from Earth's
surface

W, weight of table
gravitational pull of Earth

Newton's first law

NEWTON'S FIRST LAW

Newton's first law of motion states that 'an object continues in uniform motion in a straight line or at rest unless a resultant external force acts'. On first reading this can sound complicated but it does not really add anything to the description of a force given on page 12. All it says is that a resultant force causes acceleration. No resultant force means no acceleration – i.e 'uniform motion in a straight line'.

Book on a table at rest

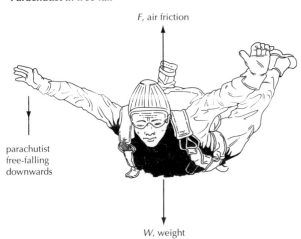

$$\text{since} \quad \text{acceleration} = \text{zero}$$
$$\text{resultant force} = \text{zero}$$
$$\therefore R - W = \text{zero}$$

Lifting a heavy suitcase

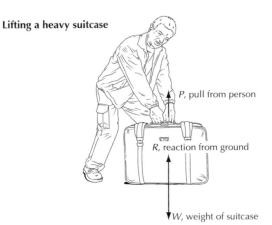

P, pull from person

R, reaction from ground

W, weight of suitcase

If the suitcase is too heavy to lift, it is not moving:
$$\therefore \text{acceleration} = \text{zero}$$
$$\therefore P + R = W$$

Parachutist in free fall

F, air friction

parachutist free-falling downwards

W, weight

If $W > F$ the parachutist accelerates downwards.
As the parachutist gets faster, the air friction increases until $W = F$
The parachutist is at constant velocity (the *acceleration* is zero).

Car travelling in a straight line

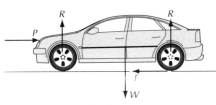

R R

P

F

W

* F is force forwards, due to engine
  P is force backwards due to air resistance

At all times force up ($2R$) = force down (W).
If $F > P$ the car accelerates forwards.
If $F = P$ the car is at constant velocity (zero acceleration).
If $F < P$ the car decelerates (i.e. there is negative acceleration and the car slows down).

Person in a lift that is moving upwards

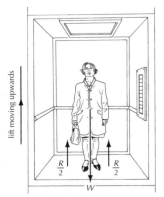

lift moving upwards

$\frac{R}{2}$ $\frac{R}{2}$

W

The total force up from the floor of the lift = R.
The total force down due to gravity = W.

If $R > W$ the person is accelerating upwards.
If $R = W$ the person is at constant velocity (acceleration = zero).
If $R < W$ the person is decelerating (acceleration is negative).

Equilibrium

EQUILIBRIUM

If the resultant force on an object is zero then it is said to be in **translational equilibrium** (or just in equilibrium). Mathematically this is expressed as follows

$$\Sigma F = zero$$

From Newton's first law, we know that the objects in the following situations must be in equilibrium.

1. An object that is constantly at rest.

2. An object that is moving with constant (uniform) velocity in a straight line.

Since forces are vector quantities, a zero resultant force means no force IN ANY DIRECTION.

For 2-dimensional problems it is sufficient to show that the forces balance in any two non-parallel directions. If this is the case then the object is in equilibrium.

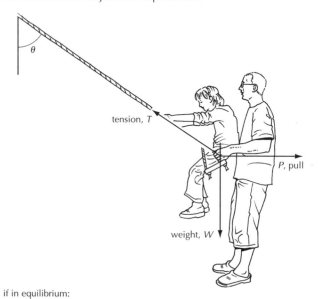

tension, T

P, pull

weight, W

if in equilibrium:
$T \sin \theta = P$ (since no resultant horizontal force)
$T \cos \theta = W$ (since no resultant vertical force)

Translational equilibrium does NOT mean the same thing as being at rest. It is even possible for an object moving with a varying velocity to pass through equilibrium. For example if the child in the previous example is allowed to swing back and forth, there are times when she is instantaneously at rest but not in equilibrium. There are also times when she is instantaneously in equilibrium, but not at rest.

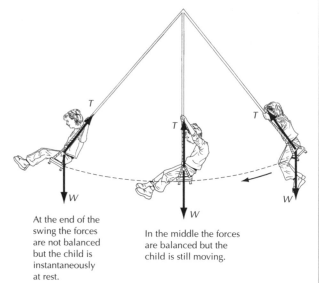

At the end of the swing the forces are not balanced but the child is instantaneously at rest.

In the middle the forces are balanced but the child is still moving.

There is another type of equilibrium called **rotational equilibrium**. This is dealt with in more detail on page 65.

DIFFERENT TYPES OF FORCES

Name of force	Description
Gravitational force	The force between objects as a results of their masses. This is sometimes referred to as the **weight** of the object but this term is, unfortunately, ambiguous – see page 17.
Electrostatic force	The force between objects as a result of their electric charges.
Magnetic force	The force between magnets and/or electric currents.
Normal reaction	The force between two surfaces that acts at right angles to the surfaces. If two surfaces are smooth then this is the only force that acts between them.
Friction	The force that opposes the relative motion of two surfaces and acts along the surfaces. **Air resistance or drag** can be thought of as a frictional force – technically this is known as **fluid friction**.
Tension	When a string (or a spring) is stretched, it has equal and opposite forces on its ends pulling outwards. The tension force is the force that the end of the string applies to another object.
Compression	When a rod is compressed (squashed), it has equal and opposite forces on its ends pushing inwards. The compression force is the force that the ends of the rod applies to another object. This is the opposite of the tension force.
Upthrust	This is the upward force that acts on an object when it is submerged in a fluid. It is the force that causes some objects to float in water.
Lift	This force can be exerted on an object when a fluid flows over it in an asymmetrical way. The shape of the wing of an aircraft causes the aerodynamic lift that enables the aircraft to fly.

Newton's second law

NEWTON'S SECOND LAW OF MOTION

Newton's first law states that a resultant force causes an acceleration. His second law provides a means of calculating the value of this acceleration. The best way of stating the second law is use the concept of the **momentum** of an object. This concept is explained on page 18.

A correct statement of Newton's second law using momentum would be 'the resultant force is proportional to the rate of change of momentum'. If we use SI units (and you always should) then the law is even easier to state – 'the resultant force is equal to the rate of change of momentum'. In symbols, this is expressed as follows

In SI units, $F = \dfrac{\Delta p}{\Delta t}$.

or, in full calculus notation, $F = \dfrac{dp}{dt}$

p is the symbol for the momentum of a body.

Until you have studied what this means this will not make much sense, but this version of the law is given here for completeness.

An equivalent (but more common) way of stating Newton's second law applies when we consider the action of a force on a single mass. If the amount of mass stays constant we can state the law as follows. 'The resultant force is proportional to the acceleration.' If we also use SI units then 'the resultant force is equal to the product of the mass and the acceleration'.

In symbols, in SI units,

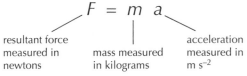

$$F = m\ a$$

resultant force measured in newtons | mass measured in kilograms | acceleration measured in m s^{-2}

Note:
- The '$F=ma$' version of the law only applies if we use SI units – for the equation to work the mass must be in **kilograms** rather than in grams.
- F is the resultant force. If there are several forces acting on an object (and this is usually true) then one needs to work out the resultant force before applying the law.
- This is an experimental law.
- There are no exceptions – Newton's laws apply throughout the Universe. (To be absolutely precise, Einstein's theory of relativity takes over at very large values of speed and mass. See page 143.)

The $F = ma$ version of the law can be used whenever the situation is simple – for example, a constant force acting on a constant mass giving a constant acceleration. If the situation is more difficult (e.g. a changing force or a changing mass) then one needs to use the

$F = \dfrac{dp}{dt}$ version.

EXAMPLES OF NEWTON'S SECOND LAW

1. Use of $F = ma$ in a simple situation

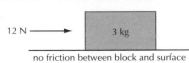

no friction between block and surface

If a mass of 3 kg is accelerated in a straight line by a resultant force of 12 N, the acceleration must be 4 m s^{-2}. Since

$F = m\,a$,

$a = \dfrac{F}{m} = \dfrac{12}{3} = 4$ m s^{-2}.

2. Use of $F = ma$ in a slightly more complicated situation

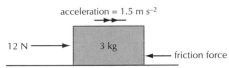

If a mass of 3 kg is accelerated in a straight line by a force of 12 N, and the resultant acceleration is 1.5 m s^{-2}, then we can work out the friction that must have been acting. Since

$$F = ma$$
$$\text{resultant force} = 3 \times 1.5$$
$$= 4.5 \text{ N}$$

This resultant force = forward force – friction

therefore, friction = forward force – resultant force
$$= 12 - 4.5 \text{ N}$$
$$= 7.5 \text{ N}$$

3. Use of $F = ma$ in a 2-dimensional situation

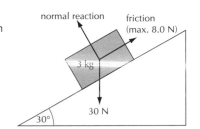

A mass of 3 kg feels a gravitational pull towards the Earth of 30 N. What will happened if it is placed on a 30 degree slope given that the maximum friction between the block and the slope is 8.0 N?

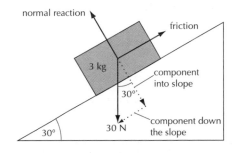

into slope: normal reaction = component into slope
The block does not accelerate into the slope.

down the slope:
component down slope = 30 N × sin 30°
$$= 15 \text{ N}$$
maximum friction force up slope = 8 N
∴ resultant force down slope = 15 – 8
$$= 7 \text{ N}$$
$$F = ma$$
∴ acceleration down slope $= \dfrac{F}{m}$
$$= \dfrac{7}{3} = 2.3 \text{ m s}^{-2}$$

Newton's third law

STATEMENT OF THE LAW

Newton's second law is an experimental law that allows us to calculate the effect that a force has. Newton's third law highlights the fact that forces always come in pairs. It provides a way of checking to see if we have remembered all the forces involved.

It is very easy to state. 'When two bodies A and B interact, the force that A exerts on B is equal and opposite to the force that B exerts on A'. Another way of saying the same thing is that 'for every action on one object there is an equal but opposite reaction on another object'.

In symbols,

$$F_{AB} = -F_{BA}$$

Key points to notice include

- the two forces in the pair act on DIFFERENT objects – this means that equal and opposite forces that act on the same object are NOT Newton's third law pairs.
- not only are the forces equal and opposite, but they must be of the same type. In other words, if the force that A exerts on B is a gravitational force, then the equal and opposite force exerted by B on A is also a gravitational force.

EXAMPLES OF THE LAW

Forces between roller-skaters

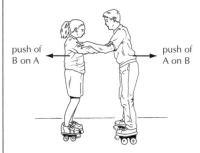

push of B on A   push of A on B

If one roller-skater pushes another, they both feel a force. The forces must be equal and opposite, but the acceleration will be different (since they have different masses).

2.0 m s⁻¹   1.5 m s⁻¹

The person with the smaller mass will gain the greater velocity.

A roller-skater pushes off from a wall

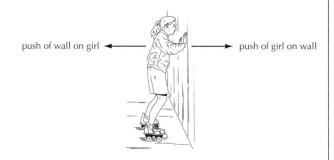

push of wall on girl ←   → push of girl on wall

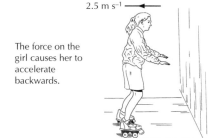

2.5 m s⁻¹ ←

The force on the girl causes her to accelerate backwards.

The mass of the wall (and Earth) is so large that the force on it does not effectively cause any acceleration.

A book on a table – Newton's third law

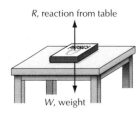

R, reaction from table

W, weight

These two forces are *not* third law pairs. There must be another force (on a different object) that pairs with each one:

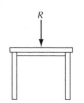

R

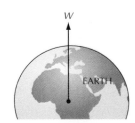

W

EARTH

If the table pushes upwards on the book with force R, then the book must push down on the table with force R.

If the Earth pulls the book down with force W, then the book must pull the Earth up with force W.

An accelerating car

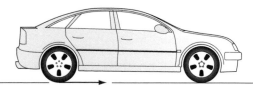

F, push forward from the ground

In order to accelerate, there must be a forward force *on the car*. The engine makes the wheels turn and the wheels push on the ground.

force from car on ground = – force from ground on car

Inertial mass, gravitational mass and weight

MASS

When we measure the mass of an object we give the result in kilograms, but what is it that we are actually measuring? There are, in fact, two very different properties measured by the one quantity that we call mass. To understand what this means, it will be useful to concentrate on each property separately and give them different names – **inertial mass** and **gravitational mass**

1. Inertial mass $m_{(i)}$

Inertial mass is the property of an object that determines how it responds to a given force – whatever the nature of the force.

A push of 700 N on a car of mass 1400 kg gives it an acceleration of 0.5 m s⁻².

The same push on a toy car would give it a much greater acceleration!

Inertial mass – different masses have different accelerations when a force acts on them

Newton's second law has already given us a rule for this ($F = ma$). This can be used to describe inertial mass. 'Inertial mass is the ratio of resultant force to acceleration'.

$$\text{Inertial mass } m_{(i)} = \frac{F}{a}$$

The units of inertial mass would be N kg⁻¹.

2. Gravitational mass $m_{(g)}$

Gravitational mass is the property of an object that determines how much gravitational force it feels when near another object.

$$\text{Gravitational force} \propto m_{(g)}$$

It also determines the gravitational force that the other object feels.

The pull of gravity (from the Earth) on an elephant is large.

The pull of gravity (from the Earth) on a mouse is much less.

Gravitational mass – different masses have different gravitational forces acting between them

As well as developing his laws of motion, Newton also developed a law of gravitational attraction. See page 61 for details.

These two concepts (inertial mass and gravitational mass) are very different. The surprise is that they turn out to be the equivalent. In other words, an object's gravitational mass is equal to its inertial mass. The fact that different objects have the same value for free-fall acceleration shows this.

A mouse and an elephant would fall together (if air friction were negligible)

WEIGHT

Mass and **weight** are two very different things. Unfortunately their meanings have become muddled in everyday language. Mass is the amount of matter contained in an object (measured in kg) whereas the weight of an object is a force (measured in N).

If an object is taken to the moon, its mass would be the same, but its weight would be less (the gravitational forces on the moon are less than on the Earth). On the Earth the two terms are often muddled because they are proportional. People talk about wanting to gain or lose weight – what they are actually worried about is gaining or losing mass.

weight, W

new weight = $2W$

Double the mass means double the weight

To make things worse, the term 'weight' can be ambiguous even to physicists. Some people choose to define weight as the gravitational force on an object. Other people define it to be the reading on a supporting scale. Whichever definition you use, you weigh less at the top of a building compared with at the bottom – the pull of gravity is slightly less!

situation:

Weight can be defined as either
(a) the pull of gravity, W or
(b) the force on a supporting scale R.

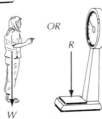

OR

Two different definitions of 'weight'

Although these two definitions are the same if the object is in equilibrium, they are very different in non-equilibrium situations. For example, if both the object and the scale were put into a lift and the lift accelerated upwards then the definitions would give different values.

If the lift is accelerating upwards:
$R > W$

The safe thing to do is to avoid using the term weight if at all possible! Stick to the phrase 'gravitational force' and you cannot go wrong.

$$\text{Gravitational force} = m\,g$$

On the surface of the Earth, g is approximately 10 N kg⁻¹, whereas on the surface of the moon, $g \approx 1.6$ N kg⁻¹

Momentum

DEFINITIONS – LINEAR MOMENTUM AND IMPULSE

Linear momentum (always given the symbol p) is defined as the product of mass and velocity.

Momentum = mass × velocity

$$p = m\,v$$

The SI units for momentum must be kg m s^{-1}. Alternative units of N s can also be used (see below). Since velocity is a vector, momentum must be a vector. In any situation, particularly if it happens quickly, the change of momentum Δp is called the **impulse** ($\Delta p = F\,\Delta t$).

USE OF MOMENTUM IN NEWTON'S SECOND LAW

Newton's second law states that the resultant force is proportional to the rate of change of momentum. Mathematically we can write this as

$$F = \frac{(final\ momentum - initial\ momentum)}{time\ taken} = \frac{\Delta p}{\Delta t}$$

Example 1

A jet of water leaves a hose and hits a wall where its velocity is brought to rest. If the hose cross-sectional area is 25 cm^2, the velocity of the water is 50 m s^{-1} and the density of the water is 1000 kg m^{-3}, what is the force acting on the wall?

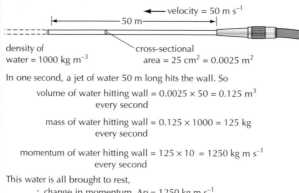

density of water = 1000 kg m^{-3}
cross-sectional area = 25 cm^2 = 0.0025 m^2

In one second, a jet of water 50 m long hits the wall. So

volume of water hitting wall = 0.0025 × 50 = 0.125 m^3 every second

mass of water hitting wall = 0.125 × 1000 = 125 kg every second

momentum of water hitting wall = 125 × 10 = 1250 kg m s^{-1} every second

This water is all brought to rest,

∴ change in momentum, Δp = 1250 kg m s^{-1}

∴ force $= \frac{\Delta p}{\Delta t} = \frac{1250}{1} = 1250$ N

Example 2

The graph below shows the variation with time of the force on a football of mass 500g. Calculate the final velocity of the ball.

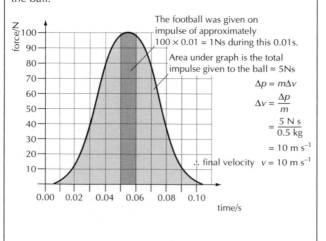

The football was given on impulse of approximately 100 × 0.01 = 1Ns during this 0.01s.

Area under graph is the total impulse given to the ball ≈ 5Ns

$\Delta p = m\Delta v$

$\Delta v = \frac{\Delta p}{m}$

$= \frac{5\ N\,s}{0.5\ kg}$

$= 10$ m s^{-1}

∴ final velocity $v = 10$ m s^{-1}

CONSERVATION OF MOMENTUM

The law of conservation of linear momentum states that 'the total linear momentum of a system of interacting particles remains constant **provided there is no resultant external force'**.

To see why, we start by imagining two isolated particles A and B that collide with one another.

- the force from A onto B, F_{AB} will cause B's momentum to change by a certain amount.
- if the time taken was Δt, then the momentum change (the impulse) given to B will be given by $\Delta p_B = F_{AB}\,\Delta t$
- by Newton's third law, the force from B onto A, F_{BA} will be equal and opposite to the force from A onto B, $F_{AB} = -F_{BA}$.
- since the time of contact for A and B is the same, then the momentum change for A is equal and opposite to the momentum change for B, $\Delta p_A = -F_{AB}\,\Delta t$.
- this means that the total momentum (momentum of A plus the momentum of B) will remain the same. Total momentum is conserved.

This argument can be extended up to any number of interacting particles so long as the system of particles is still isolated. If this is the case, the momentum is still conserved.

ELASTIC AND INELASTIC COLLISIONS

The law of conservation of linear momentum is not enough to always predict the outcome after a collision (or an explosion). This depends on the nature of the colliding bodies. For example, a moving railway truck, m_A, velocity v, collides with an identical stationary truck m_B. Possible outcomes are:

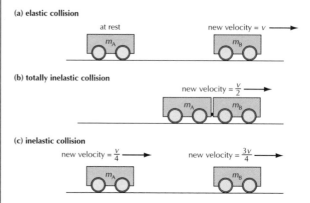

In (a), the trucks would have to have elastic bumpers. If this were the case then no mechanical energy at all would be lost in the collision. A collision in which no mechanical energy is lost is called an **elastic collision**. In reality, collisions between everyday objects always lose some energy – the only real example of elastic collisions is the collision between molecules. For an elastic collision, the relative velocity of approach always equals the relative velocity of separation.

In (b), the railway trucks stick together during the collision (the relative velocity of separation is zero). This collision is what is known as a **totally inelastic collision**. A large amount of mechanical energy is lost (as heat and sound), but the total momentum is still conserved.

In energy terms, (c) is somewhere between (a) and (b). Some energy is lost, but the railway trucks do not join together. This is an example of an **inelastic collision**. Once again the total momentum is conserved.

Work

WHEN IS WORK DONE?

Work is done when a force moves its point of application in the direction of the force. If the force moves at right angles to the direction of the force, then no work has been done.

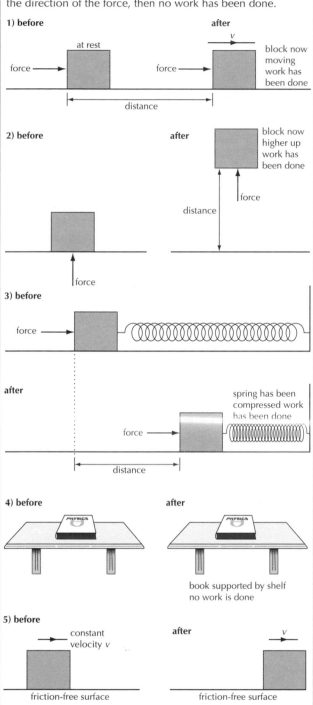

1) before   **after**

at rest

force →   force →

block now moving work has been done

distance

2) before   **after**

block now higher up work has been done

distance

force ↑

force ↑

3) before

force →

after

spring has been compressed work has been done

force →

distance

4) before   **after**

PHYSICS

PHYSICS

book supported by shelf no work is done

5) before   **after**

constant velocity v →

v →

friction-free surface

friction-free surface

object continues at constant velocity no work is done

In the examples above the work done has had different results.

- In 1 the force has made the object move faster.
- In 2 the object has been lifted higher in the gravitational field.
- In 3 the spring has been compressed.
- In 4 and 5, NO work is done. Note that even though the object is moving in the last example, there is no force moving along its direction of action so no work is done.

DEFINITION OF WORK

Work is a scalar quantity. Its definition is as follows.

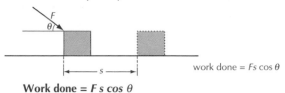

work done = $Fs \cos \theta$

Work done = $F s \cos \theta$

If the force and the displacement are in the same direction, this can be simplified to

'Work done = force × distance'

From this definition, the SI units for work done are N m. We define a new unit called the joule: 1 J = 1 N m.

EXAMPLES

(1) lifting vertically

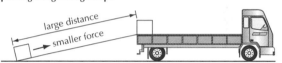

small distance

large force

(2) pushing along a rough slope

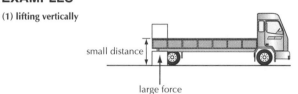

large distance

smaller force

The task in the second case would be easier to perform (it involves less force) but overall it takes more work since work has to be done to overcome friction. In each case, the useful work is the same.

If the force doing work is not constant (for example, when a spring is compressed), then graphical techniques can be used.

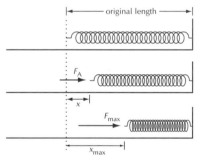

original length

F_A

x

F_{max}

x_{max}

The total work done is the area under the force-displacement graph.

force

F_{max}

total work done = area under graph

0   x_{max} extension

Useful equations for the work done include

- work done when lifting something vertically = mgh
  where m represents mass (in kg)
  g represents the Earth's gravitational field strength (10 N kg^{-1}) h represents the height change (in m)

- work done in compressing or extending a spring = $\frac{1}{2} kx^2$

Energy and power

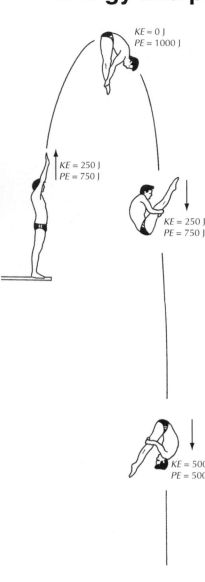

KE ≈ 0 J
PE = 1000 J

KE = 250 J
PE = 750 J

KE = 250 J
PE = 750 J

KE = 500 J
PE = 500 J

KE = 750 J
PE = 250 J

KE = 1000 J
PE = 0 J

CONCEPTS OF ENERGY AND WORK

Energy and work are linked together. When you do work on an object, it gains energy and you lose energy. **The amount of energy transferred is equal to the work done**. Energy is a measure of the amount of work done. This means that the units of energy must be the same as the units of work – joules.

ENERGY TRANSFORMATIONS – CONSERVATION OF ENERGY

In any situation, we must be able to account for the changes in energy. If it is 'lost' by one object, it must be gained by another. This is known as the **principle of conservation of energy**. There are several ways of stating this principle:

- Overall the total energy of any closed system must be constant.
- Energy is neither created or destroyed, it just changes form.
- There is no change in the total energy in the Universe.

ENERGY TYPES

Kinetic energy	**Electrostatic potential**	**Chemical energy**	**Radiant energy**
Gravitational potential	**Thermal energy**	**Nuclear energy**	**Solar energy**
Elastic potential energy	**Electrical energy**	**Internal energy**	**Light energy**

Equations for the first three types of energy are given below.

Kinetic energy = $\frac{1}{2} mv^2$ where m is the mass (in kg), v is the velocity (in m s^{-1})

Gravitational potential energy = mgh where m represents mass (in kg), g represents the Earth's gravitational field (10 N kg^{-1}), h represents the height change (in m)

Elastic potential energy = $\frac{1}{2} kx^2$ where k is the spring constant (in N m^{-1}), x is the extension (in m)

EFFICIENCY AND POWER

1. Power

Power is defined as the RATE at which energy is transferred. This is the same as the rate at which work is done.

$$\text{Power} = \frac{\text{energy transferred}}{\text{time taken}} = \frac{\text{work done}}{\text{time taken}}$$

The SI unit for power is the joule per second (J s^{-1}). Another unit for power is defined – the watt (W). 1 W = 1 J s^{-1}.

If something is moving at a constant velocity v against a constant frictional force f, the power P needed is $P = f v$

2. Efficiency

Depending on the situation, we can categorise the energy transferred (work done) as useful or not. In a light bulb, the useful energy would light energy, the 'wasted' energy would be thermal energy (and non-visible forms of radiant energy).

We define efficiency as the ratio of useful energy to the total energy transferred. Possible forms of the equation include:

$$\text{Efficiency} = \frac{\text{useful work OUT}}{\text{total energy transformed}}$$

$$\text{Efficiency} = \frac{\text{useful energy OUT}}{\text{total energy IN}}$$

$$\text{Efficiency} = \frac{\text{useful power OUT}}{\text{total power IN}}$$

Since this is a ratio it does not have any units. Often it is expressed as a percentage.

EXAMPLE

A grasshopper (mass 8g) uses its hindlegs to push for 0.1s and as a result jumps 1.8m high. Calculate (i) its take off speed, (ii) the power developed.

(i) P.E. gained = mgh

K.E at start = $\frac{1}{2} mv^2$

$\frac{1}{2} mv^2 = mgh$ (conservation of energy)

$v = \sqrt{2gh} = \sqrt{2 \times 10 \times 1.8} = 6$ ms^{-1}

(ii) Power = $\frac{mgh}{t}$

$= \frac{0.008 \times 10 \times 1.8}{0.1}$

≈ 1.4 W

Uniform circular motion

MECHANICS OF CIRCULAR MOTION

The phrase 'uniform circular motion' is used to describe an object that is going around a circle at constant speed. Most of the time this also means that the circle is horizontal. An example of uniform circular motion would be the motion of a small mass on the end of a string as shown below.

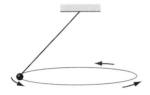

mass moves at constant speed

Example of uniform circular motion

It is important to remember that even though the speed of the object is constant, its direction is changing all the time.

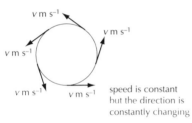

speed is constant but the direction is constantly changing

Circular motion. The direction of motion is changing all the time.

This constantly changing direction means that the velocity of the object is constantly changing. The word 'acceleration' is used whenever an object's velocity changes. This means that an object in uniform circular motion MUST be accelerating even if the speed is constant.

The acceleration of a particle travelling in circular motion is called the **centripetal acceleration**. The force needed to cause the centripetal acceleration is called the **centripetal force**.

MATHEMATICS OF CIRCULAR MOTION

The diagram below allows us to work out the direction of the centripetal acceleration – which must also be the direction of the centripetal force. This direction is constantly changing.

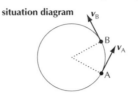 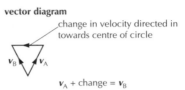

v_A + change = v_B

The object is shown moving between two points A and B on a horizontal circle. Its velocity has changed from v_A to v_B. The magnitude of velocity is always the same, but the direction has changed. Since velocities are vector quantities we need to use vector mathematics to work out the average change in velocity. This vector diagram is also shown above.

In this example, the direction of the average change in velocity is towards the centre of the circle. This is always the case and thus true for the instantaneous acceleration. For a mass m moving at a speed v in uniform circular motion of radius r,

Centripetal acceleration $a_{centripetal} = \dfrac{v^2}{r}$ [In towards the centre of the circle]

A force must have caused this acceleration. The value of the force is worked out using Newton's second law:

Centripetal force (CPF) $f_{centripetal} = m\, a_{centripetal}$

$$= \dfrac{m v^2}{r} \text{ [In towards the centre of the circle]}$$

For example, if a car of mass 1500 kg is travelling at a constant speed of 20 m s^{-1} around a circular track of radius 50 m, the resultant force that must be acting on it works out to be

$$F = \dfrac{1500\,(20)^2}{50} = 12\,000 \text{ N}$$

It is really important to understand that centripetal force is NOT a new force that starts acting on something when it goes in a circle. It is a way of working out what the total force must have been. This total force must result from all the other forces on the object. See the examples below for more details.

One final point to note is that the centripetal force does NOT do any work. (Work done = force × distance **in the direction of the force**.)

EXAMPLES

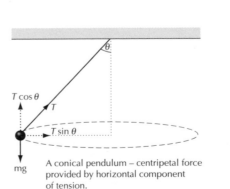 Earth's gravitational attraction on moon. Moon. Earth

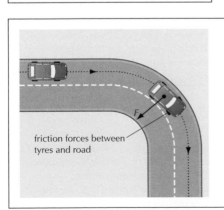

friction forces between tyres and road

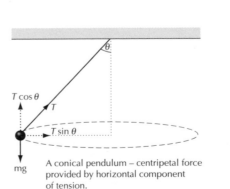

$T \cos\theta$, T, $T \sin\theta$, mg

A conical pendulum – centripetal force provided by horizontal component of tension.

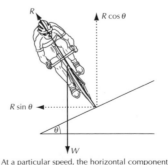

R, $R\cos\theta$, $R\sin\theta$, θ, W

At a particular speed, the horizontal component of the normal reaction can provide all the centripetal force (without needing friction).

IB QUESTIONS – MECHANICS

1 Two identical objects A and B fall from rest from different heights. If B takes twice as long as A to reach the ground, what is the ratio of the heights from which A and B fell? Neglect air resistance.

   A 1:√2　　**B** 1:2　　**C** 1:4　　**D** 1:8

2 A trolley is given an initial push along a horizontal floor to get it moving. The trolley then travels forward along the floor, gradually slowing. What is true of the horizontal force(s) on the trolley while it is slowing?

   A There is a forward force and a backward force, but the forward force is larger.

   B There is a forward force and a backward force, but the backward force is larger.

   C There is only a forward force, which diminishes with time.

   D There is only a backward force.

3 A mass is suspended by cord from a ring which is attached by two further cords to the ceiling and the wall as shown. The cord from the ceiling makes an angle of less than 45° with the vertical as shown. The tensions in the three cords are labelled R, S and T in the diagram.

   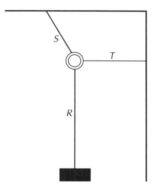

   How do the tensions R, S and T in the three cords compare in magnitude?

   A $R > T > S$　　　**B** $S > R > T$

   C $R = S = T$　　　**D** $R = S > T$

4 In any collision between two objects, what is true about the total momentum and the total kinetic energy of the system of two objects?

	Total momentum	Total kinetic energy
A	always stays the same	always stays the same
B	always stays the same	can change
C	can change	always stays the same
D	can change	can change

5 A ball is dropped on to a hard surface and makes several bounces before coming to rest. Which one of the graphs below best represents how the velocity of the ball varies with time?

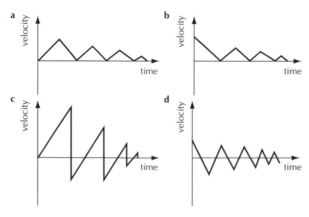

6 A car and a truck are both travelling at the speed limit of 60 km h⁻¹ but in opposite directions as shown. The truck has **twice** the mass of the car.

The vehicles collide head-on and become entangled together.

   (a) During the collision, how does the force exerted by the car on the truck compare with the force exerted by the truck on the car? Explain. [2]

   (b) In what direction will the entangled vehicles move after collision, or will they be stationary? Support your answer, referring to a physics principle. [2]

   (c) Determine the speed (in km h⁻¹) of the combined wreck immediately after the collision. [3]

   (d) How does the acceleration of the car compare with the acceleration of the truck during the collision? Explain. [2]

   (e) Both the car and truck drivers are wearing seat belts. Which driver is likely to be more severely jolted in the collision? Explain. [2]

   (f) The total kinetic energy of the system decreases as a result of the collision. Is the principle of conservation of energy violated? Explain. [1]

7 A car travels at a steady speed v in a circular path of radius r on a circular track banked at an angle θ, as shown in the plan and side views in the diagram.

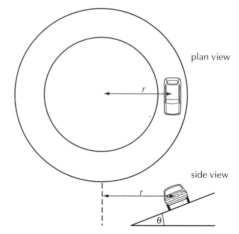

The car's speed is such that there is no sideways frictional force between the tyres and the track.

   (a) Does the car have an acceleration? Explain why or why not. If you say yes, state its direction. [2]

   (b) The car on the track is represented by a block in the figure below, moving perpendicular to the page. Draw a force diagram, showing and labelling all the forces acting on the moving car. [2]

   (c) The track is banked at an angle of 17° and the circular path of the car has radius 30 m. Calculate the speed at which the car must travel in order that there be no sideways frictional force between the tyres and the track. Show all working. [4]

Temperature and thermometers

TEMPERATURE AND HEAT FLOW

Hot and cold are just labels that identify the direction in which thermal energy (otherwise known as heat) will be naturally transferred when two objects are placed in thermal contact. This leads to the concept of the 'hotness' of an object. The direction of the natural flow of thermal energy between two objects is determined by the 'hotness' of each object. Thermal energy naturally flows from hot to cold.

The temperature of an object is a measure of how hot it is. In other words, if two objects are placed in thermal contact, then the temperature difference between the two objects will determine the direction of the natural transfer of thermal energy. Thermal energy is naturally transferred 'down' the temperature difference – from high temperature to low temperature. Eventually, the two objects would be expected to reach the same temperature. When this happens, they are said to be in **thermal equilibrium**.

Heat is not a substance that flows from one object to another. What has happened is that thermal energy has been transferred. The historical perspective on this is studied in option E (see page 113).

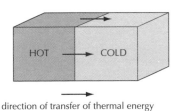

direction of transfer of thermal energy

SCALES OF TEMPERATURE

There are many possible scales of temperature that can be used.

To define a scale of temperature we need
• a property that varies with temperature, and
• a way of defining two fixed points on the scale (the 'lower fixed point' and the 'higher fixed point').

In outline, the procedure is as follows
• the value of the property is measured at both fixed points.
• the temperature of any intermediate point is found by interpolation.
• the temperature of any point outside the two fixed points is found by extrapolation.

For example, a property that varies with temperature could be the length of a piece of metal. As the temperature goes up, the metal expands. As the temperature goes down, the metal contracts. The lower fixed point could be chosen as the freezing point of water and defined to be zero on the scale. The higher fixed point could be chosen as the boiling point of water and defined to be 100 units on the scale.

The length of this metal can be used to define a temperature scale and can be used to measure an unknown temperature. If the length of the metal at some unknown temperature is half way between the length it has at the lower fixed point and the length it has at the upper fixed point, then the temperature must be half way between zero and 100, i.e. 50.

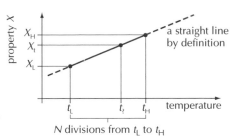

The definition of a temperature in terms of a property x.

If the property being considered is X, this can be expressed as follows …

X_H is the value of the property at the higher fixed point, t_H

X_L is the value of the property at the lower fixed point, t_L

X_t is the value of the property at the unknown temperature, $t_?$

N is the number of divisions (the number of degrees) that we chose to have between the fixed points.

Then the temperature $t_?$ is equal to $\dfrac{(X_t - X_L)}{(X_H - X_L)} N + t_L$

KELVIN AND CELSIUS

Most of the time, there are only two sensible temperature scales to chose between – the Kelvin scale and the Celsius scale.

In order to use them, you do not need to understand the details of how either of these scales has been defined, but you do need to know the relation between them. Most everyday thermometers are marked with the Celsius scale – temperature is quoted in degrees Celsius (°C).

There is an easy relationship between a temperature T as measured on the Kelvin scale and the corresponding temperature t as measured on the Celsius scale. The approximate relationship is

T (K) = t (°C) + 273

This means that the 'size' of the units used on each scale is identical, but they have different zero points.

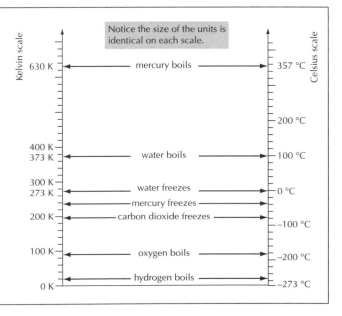

Heat and internal energy

MICROSCOPIC VS MACROSCOPIC

When analysing something physical, we have a choice.

- the **macroscopic** point of view considers the system as a whole and sees how it interacts with its surroundings.
- the **microscopic** point of view looks inside the system to see how its component parts interact with each other.

So far we have looked at the temperature of an system in a macroscopic way, but all objects are made up of **atoms** and **molecules**.

According to **kinetic theory** these particles are constantly in random motion – hence the name. See below for more details. Although atoms and molecules are different things (a molecule is a combination of atoms), the difference is not important at this stage. The particles can be thought of as little 'points' of mass with velocities that are continually changing.

INTERNAL ENERGY

If the temperature of an object changes then it must have gained (or lost) energy. From the microscopic point of view, the molecules must have gained (or lost) this energy.

The two possible forms are kinetic energy and potential energy.

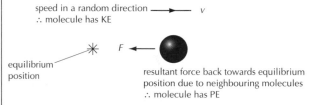

- The molecules have kinetic energy because they are moving. To be absolutely precise, a molecule can have either translational kinetic energy (the whole molecule is moving in a certain direction) or rotational kinetic energy (the molecule is rotating about one or more axes).
- The molecules have potential energy because of the **intermolecular** forces. If we imagine pulling two molecules further apart, this would require work against the intermolecular forces.

The total energy that the molecules possess (kinetic plus potential) is called the **internal energy** of a substance. Whenever we heat a substance, we increase its internal energy.

Temperature is a measure of the average kinetic energy of the molecules in a substance.

If two substances have the same temperature, then their molecules have the same average kinetic energy.

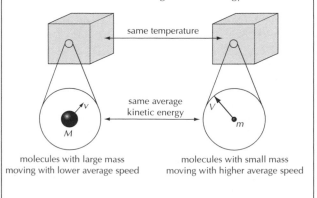

molecules with large mass moving with lower average speed

molecules with small mass moving with higher average speed

KINETIC THEORY

Molecules are arranged in different ways depending on the **phase** of the substance (i.e. solid, liquid or gas).

Solids

Macroscopically, solids have a fixed volume and a fixed shape. This is because the molecules are held in position by bonds. However the bonds are not absolutely rigid. The molecules vibrate around a mean (average) position. The higher the temperature, the greater the vibrations.

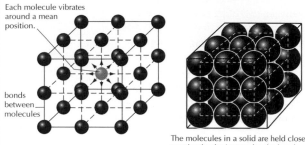

Each molecule vibrates around a mean position.

bonds between molecules

The molecules in a solid are held close together by the intermolecular bonds.

Liquids

A liquid also has a fixed volume but its shape can change. The molecules are also vibrating, but they are not completely fixed in position. There are still strong forces between the molecules. This keeps the molecules close to one another, but they are free to move around each other.

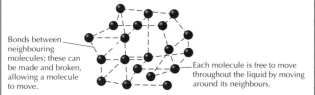

Bonds between neighbouring molecules; these can be made and broken, allowing a molecule to move.

Each molecule is free to move throughout the liquid by moving around its neighbours.

Gases

A gas will always expand to fill the container in which it is put. The molecules are not fixed in position, and any forces between the molecules are very weak. This means that the molecules are essentially independent of one another, but they do occasionally collide. More detail is given on page 29.

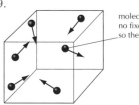

molecules in random motion no fixed bonds between molecules so they are free to move

HEAT AND WORK

Many people have confused ideas about heat and work. In answers to examination questions it is very common to read, for example, that 'heat rises' – when what is meant is that the transfer of thermal energy is upwards.

- When a force moves through a distance, we say that work is done. Work is the energy that has been transmitted from one system to another from the macroscopic point of view.
- When work is done on a microscopic level (i.e. on individual molecules), we say that heating has taken place. Heat is the energy that has been transmitted. It can either increase the kinetic energy of the molecules or their potential energy or, of course, both.

In both cases energy is being transferred.

Thermal energy transfer

PROCESSES OF THERMAL ENERGY TRANSFER

There are several processes by which the transfer of thermal energy from a hot object to a cold object can be achieved. Three very important processes are called **conduction**, **convection** and **radiation**. Any given practical situation probably involves more than one of these processes happening at the same time. There is a fourth process called **evaporation**, the details of which are shown on page 27.

CONDUCTION

In thermal conduction, thermal energy is transferred along a substance without any bulk (overall) movement of the substance. For example, one end of a metal spoon soon feels hot if the other end is placed in a hot cup of tea.

Conduction is the process by which kinetic energy is passed from molecule to molecule.

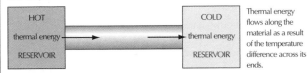
macroscopic view

Thermal energy flows along the material as a result of the temperature difference across its ends.

microscopic view

The faster-moving molecules at the hot end pass on their kinetic energy to the slower-moving molecules as a result of intermolecular collisions.

Points to note:
- poor conductors are called **insulators**.
- metals tend to be very good thermal conductors. This is because a different mechanism (involving the electrons) allows quick transfer of thermal energy.
- all gases (and most liquids) tend to be poor conductors.

Examples
- Most clothes keep us warm by trapping layers of air – a poor conductor
- If one walks around a house in bare feet, the floors that are better conductors (e.g. tiles) will feel colder than the floors that are good insulators (e.g. carpets) even if they are at the same temperature. (For the same reason, on a cold day a piece of metal feels colder than piece of wood.)
- When used for cooking food, saucepans conduct thermal energy from the source of heat to the food.

EXAMPLE

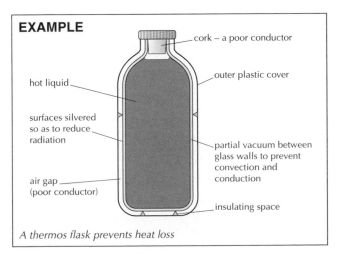

cork – a poor conductor

outer plastic cover

hot liquid

surfaces silvered so as to reduce radiation

partial vacuum between glass walls to prevent convection and conduction

air gap (poor conductor)

insulating space

A thermos flask prevents heat loss

CONVECTION

In convection, thermal energy moves between two points because of a bulk movement of matter. This can only take place in a **fluid** (a liquid or a gas). When part of the fluid is heated it tends to expand and thus its density is reduced. The colder fluid sinks and the hotter fluid rises up. Central heating causes a room to warm up because a **convection current** is set up as shown below.

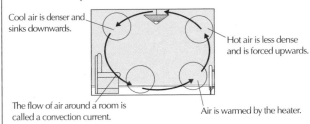

Cool air is denser and sinks downwards.

Hot air is less dense and is forced upwards.

The flow of air around a room is called a convection current.

Air is warmed by the heater.

Convection in a room

Points to note:
- convection cannot take place in a solid.

Examples
- The pilots of gliders (and many birds) use naturally occurring convection currents in order to stay above the ground.
- Sea breezes (winds) are often due to convection. During the day the land is hotter than the sea. This means hot air will rise from above the land and there will be a breeze onto the shore. During the night, the situation is reversed.
- Lighting a fire in a chimney will mean that a breeze flows in the room towards the fire.

RADIATION

Matter is not involved in the transfer of thermal energy by radiation. All objects (that have a temperature above zero kelvin) radiate **electromagnetic waves**. If you hold your hand up to a fire to 'feel the heat', your hands are receiving the radiation.

For most everyday objects this radiation is in the **infra-red** part of the **electromagnetic spectrum**. For more details of the electromagnetic spectrum – see page 157.

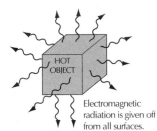

HOT OBJECT

Electromagnetic radiation is given off from all surfaces.

Points to note:
- an object at room temperature absorbs and radiates energy. If it is at constant temperature (and not changing state) then the rates are the same.
- a surface that is a good radiator is also a good absorber.
- surfaces that are light in colour and smooth (shiny) are poor radiators (and poor absorbers).
- surfaces that are dark and rough are good radiators (and good absorbers).
- if the temperature of an object is increased then the frequency of the radiation increases. The total rate at which energy is radiated will also increase.
- radiation can travel through a vacuum (space).

Examples
- The Sun warms the Earth's surface by radiation.
- Clothes in summer tend to be white – so as not to absorb the radiation from the Sun.

Specific heat capacity

DEFINITIONS & MICROSCOPIC EXPLANATION

In theory, if an object could be heated up with no energy loss, then the increase in temperature $\Delta\theta$ depends on three things:

- the energy given to the object ΔQ,
- the mass, m, and,
- the substance from which the object is made.

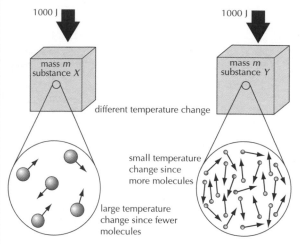

Two different blocks with the same mass and same energy input will have a different temperature change.

We define the **heat capacity** C of an object as the energy required to raise its temperature by 1 K. Different objects (even different samples of the same substance) will have different values of heat capacity. **Specific heat capacity** is the energy required to raise a unit mass of a substance by 1 K. 'Specific' here just means 'per unit mass'.

In symbols,

> **Heat capacity** $\quad C = \dfrac{\Delta Q}{\Delta T}$ (J K⁻¹ or J °C⁻¹)

> **Specific heat capacity** $\quad c = \dfrac{\Delta Q}{(m\,\Delta T)}$ (J kg⁻¹ K⁻¹ or J kg⁻¹ °C⁻¹)

Note

- a particular gas can have many different values of specific heat capacity – it depends on the conditions used – see page 68.
- these equations refer to the **temperature difference** resulting from the addition of a certain amount of energy. In other words, it generally takes the same amount of energy to raise the temperature of an object from 25 °C to 35 °C as it does for the same object to go from 402 °C to 412 °C. This is only true so long as energy is not lost from the object.
- if an object is raised above room temperature, it starts to lose energy. The hotter it becomes, the greater the rate at which it loses energy.

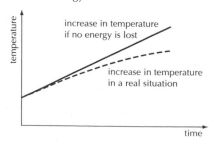

Temperature change of an object being heated at a constant rate

METHODS OF MEASURING HEAT CAPACITIES AND SPECIFIC HEAT CAPACITIES

The are two basic ways to measure heat capacity.

1. Electrical method

The experiment would be set up as below:

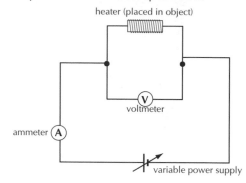

- the specific heat capacity $c = \dfrac{I\,t\,V}{m\,(T_2 - T_1)}$.

Sources of experimental error

- the loss of thermal energy from the apparatus.
- container for the substance and the heater will also be warmed up.
- it will take some time for the energy to be shared uniformly through the substance.

2. Method of mixtures

The known specific heat capacity of one substance can be used to find the specific heat capacity of another substance.

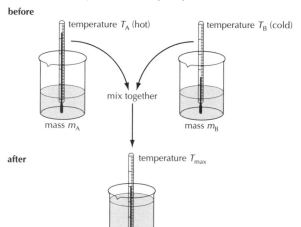

Procedure:

- measure the masses of the liquids m_A and m_B.
- measure the two starting temperatures T_A and T_B.
- mix the two liquids together.
- record the maximum temperature of the mixture T_{max}.

If no energy is lost from the system then,

> **energy lost by hot substance cooling down = energy gained by cold substance heating up**

> $m_A\,c_A\,(T_A - T_{max}) = m_B\,c_B\,(T_{max} - T_B)$

Again, the main source of experimental error is the loss of thermal energy from the apparatus – particularly while the liquids are being transferred. The changes of temperature of the container also need to be taken into consideration for a more accurate result.

Phases (states) of matter and latent heat

DEFINITIONS AND MICROSCOPIC VIEW

When a substance changes phase, the temperature remains constant even though thermal energy is still being transferred.

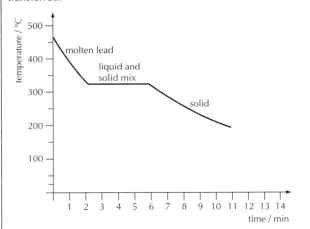

Cooling curve for molten lead (idealized)

The amount of energy associated with the phase change is called the **latent heat**. The technical term for the change of phase from solid to liquid is **fusion** and the term for the change from liquid to gas is **vaporization**.

The energy given to the molecules does not increase their kinetic energy so it must be increasing their potential energy. Intermolecular bonds are being broken and this takes energy. When the substance freezes bonds are created and this process releases energy.

It is a very common mistake to think that the molecules must speed up during a phase change. The molecules in water vapour at 100 °C must be moving with the same average speed as the molecules in liquid water at 100 °C.

The **specific latent heat** of a substance is defined as the amount of energy per unit mass absorbed or released during a change of phase.

In symbols,

Specific latent heat $l = \dfrac{\Delta Q}{m}$ $\quad$ (J kg^{-1}.)

EVAPORATION

Evaporation takes place at the surface of liquids. If a liquid is below its boiling point, on average the molecules do not have sufficient energy to leave the surface.

The result of the overall process is for the faster moving molecules to escape the liquid. This means that it is the slower moving ones that are left behind – in other words the temperature of the liquid falls as a result of evaporation. Evaporation causes cooling.

The rate at which evaporation takes place depends on
- the surface area of the liquid (increased area means increased evaporation rate).
- the temperature of the liquid (increased temperature means increased evaporation rate).
- the pressure (or moisture content) of the air above the liquid (increased pressure means decreased evaporation rate).
- any draught that exists above the liquid (increased draught means increased evaporation rate).

METHODS OF MEASURING

The two possible methods for measuring latent heats shown below are very similar in principle to the methods for measuring specific heat capacities (see previous page)

1. A method for measuring the specific latent heat of vaporisation of water

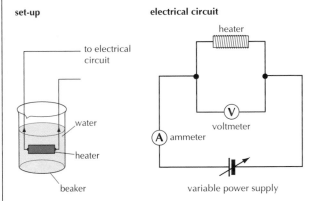

A method for measuring the latent heat of vaporization.

The amount of thermal energy provided to water at its boiling point is calculated using electrical energy = $I\,t\,V$. The mass vaporised needs to be recorded.

- the specific latent heat $l = \dfrac{I\,t\,V}{(m_1 - m_2)}$.

Sources of experimental error
- loss of thermal energy from the apparatus.
- some water vapour will be lost before and after timing.

2. A method for measuring the specific latent heat of fusion of water

Providing we know the specific heat capacity of water, we can calculate the specific latent heat of fusion for water. In the example below ice (at 0 °C) is added to warm water and the temperature of the resulting mix is measured.

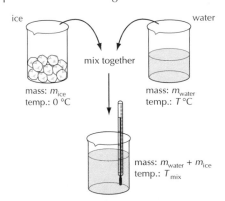

A method for measuring the latent heat of fusion.

If no energy is lost from the system then,

energy lost by water cooling down = energy gained by ice

$m_{water}\, c_{water}\, (T_{water} - T_{mix}) = m_{ice}\, l_{fusion} + m_{ice}\, c_{water}\, T_{mix}$

Sources of experimental error
- loss (or gain) of thermal energy from the apparatus.
- if the ice had not started at exactly zero, then there would be an additional term in the equation in order to account for the energy needed to warm the ice up to 0 °C.
- water clinging to the ice before the transfer.

The gas laws

GAS LAWS

For a given sample of a gas, the pressure, volume and temperature are all related to one another.

The graphs below outline what might be observed experimentally.

(a) constant volume

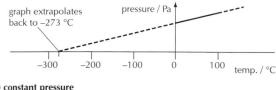

graph extrapolates back to –273 °C

(b) constant pressure

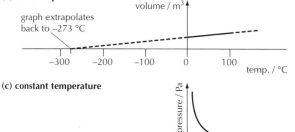

graph extrapolates back to –273 °C

(c) constant temperature

Points to note:
- Although pressure and volume both vary linearly with Celsius temperature, neither pressure nor volume is proportional to Celsius temperature.

DEFINITIONS

The concepts of the **mole**, **molar mass** and the **Avogadro constant** are all introduced so as to be able to relate the mass of a gas (an easily measurable quantity) to the number of molecules that are present in the gas.

Ideal gas	An ideal gas is one that follows the gas laws for all values of of P, V and T.
Mole	The mole is the basic SI unit for 'amount of substance'. One mole of any substance is equal to the amount of that substance that contains the same number of atoms as 0.012 kg of carbon–12 (^{12}C). When writing the unit it is (slightly) shortened to the mol.
Avogadro constant	This is the number of atoms in 0.012 kg of carbon–12 (^{12}C). It is 6.02×10^{23}.
Molar mass	The mass of one mole of a substance is called the molar mass. A simple rule applies. If an element has a certain mass number, A, then the molar mass will be A grams.

EXAMPLE

What volume will be occupied by 8g of helium (mass number 4) at room temperature (20°C) and atmospheric pressure ($1.0 \times 10^5 Nm^{-2}$)

$$n = \frac{8}{4} = 2 \text{ moles}$$

$$T = 20 + 273 = 293k$$

$$V = \frac{nRT}{P} = \frac{2 \times 8.314 \times 293}{1.0 \times 10^5} = 0.049m^3$$

- A different sample of gas would produce a different straight-line variation for pressure (or volume) against temperature but both graphs would extrapolate back to the same low temperature, –273 °C. This temperature is known as **absolute zero**.
- As pressure increases, the volume decreases. In fact they are inversely proportional.

The trends can be seen more clearly if this information is presented in a slightly different way.

(1) constant volume

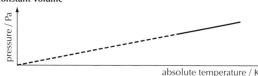

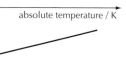

(2) constant pressure

(3) constant temperature

From these graphs for a fixed mass of gas we can say that

1. At constant V, $P \propto T$ or $\frac{P}{T}$ = constant (the pressure law)

2. At constant P, $V \propto T$ or $\frac{V}{T}$ = constant (Charles's law)

3. At constant T, $P \propto \frac{1}{V}$ or $P V$ = constant (Boyle's law)

These relationships are known as the **ideal gas laws**. The temperature is always expressed in kelvin. (These laws do not always apply to experiments done with real gases. A real gas is said to 'deviate' from ideal behaviour under certain conditions (e.g. high pressure).)

EQUATION OF STATE

The three ideal gas laws can be combined together to produce one mathematical relationship.

$$\frac{PV}{T} = \text{constant}.$$

This constant will depend on the mass and type of gas.

If we compare the value of this constant for different masses of different gases, it turns out to depend on the number of molecules that are in the gas – not their type. In this case we use the definition of the mole to state that for n moles of ideal gas

$$\frac{PV}{nT} = \text{a universal constant}.$$

The universal constant is called the **molar gas constant** R.

The SI unit for R is J mol^{-1} K^{-1}

$$R = 8.314 \text{ J mol}^{-1} \text{ K}^{-1}$$

Summary: $\frac{PV}{nT} = R$ Or $P V = n R T$

Molecular model of an ideal gas

KINETIC MODEL OF AN IDEAL GAS

Assumptions:

- Newton's laws apply to molecular behaviour
- there are no intermolecular forces
- the molecules are perfect spheres (treated as points)
- the molecules are in random motion
- the collisions between the molecules are elastic (no energy is lost)
- there is no time spent in these collisions

The pressure of a gas is explained as follows:

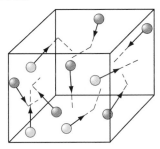

The pressure of a gas is a result of collisions between the molecules and the walls of the container.

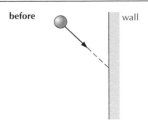

before   wall

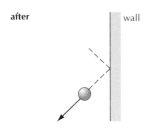

after   wall

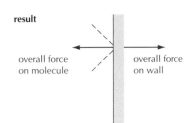

result   wall

overall force on molecule   overall force on wall

A single molecule hitting the walls of the container.

- When a molecule bounces off the walls of a container its momentum changes (due to the change in direction – momentum is a vector).
- There must have been a force on the molecule from the wall (Newton II).
- There must have been an equal and opposite force on the wall from the molecule (Newton III).
- Each time there is a collision between a molecule and the wall, a force is exerted on the wall.
- The average of all the microscopic forces on the wall over a period of time means that there is effectively a constant force on the wall from the gas.
- This force per unit area of the wall is what we call pressure.

Since the temperature of a gas is a measure of the average kinetic energy of the molecules, as we lower the temperature of a gas the molecules will move slower. At absolute zero, we imagine the molecules to have zero kinetic energy. We cannot go any lower because we cannot reduce their kinetic energy any further!

PRESSURE LAW

Macroscopically, at a constant volume the pressure of a gas is proportional to its temperature in kelvin. Microscopically this can be analysed as follows

- If the temperature of a gas goes up, the molecules have more average kinetic energy – they are moving faster on average.
- Fast moving molecules will have a greater change of momentum when they hit the walls of the container.
- Thus the microscopic force from each molecule will be greater.
- The molecules are moving faster so they hit the walls more often.
- For both these reasons, the total force on the wall goes up.
- Thus the pressure goes up.

low temperature   high temperature

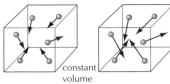

low pressure   high pressure

Microscopic justification of the pressure law

CHARLES'S LAW

Macroscopically, at a constant pressure, the volume of a gas is proportional to its temperature in kelvin. Microscopically this can be analysed as follows

- A higher temperature means faster moving molecules (see above).
- Faster moving molecules hit the walls with a greater microscopic force (see above).
- If the volume of the gas increases, then the rate at which these collisions take place on a unit area of the wall must goes down.
- The average force on a unit area of the wall can thus be the same.
- Thus the pressure remains the same.

low temperature   high temperature

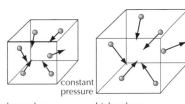

low volume   high volume

Microscopic justification of Charles's law

BOYLE'S LAW

Macroscopically, at a constant temperature, the pressure of a gas is inversely proportional to its volume. Microscopically this can be seen to be correct.

- The constant temperature of gas means that the molecules have a constant average speed.
- The microscopic force that each molecule exerts on the wall will remain constant
- Increasing the volume of the container decreases the rate with which the molecules hit the wall – average total force decreases.
- If the average total force decreases the pressure decreases.

high pressure   low pressure

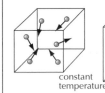
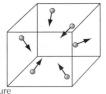

low volume   high volume

Microscopic justification of Boyle's law

IB QUESTIONS – THERMAL PHYSICS

The following information relates to questions 1 and 2 below.

A substance is heated at a constant rate of energy transfer. A graph of its temperature against time is shown below.

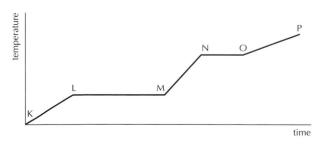

1 Which regions of the graph correspond to the substance existing in a mixture of two phases?

   A KL, MN and OP   **C** All regions

   B LM and NO   **D** No regions

2 In which region of the graph is the specific heat capacity of the substance greatest?

   A KL   **C** MN

   B LM   **D** OP

3 When the volume of a gas is isothermally compressed to a smaller volume, the pressure exerted by the gas on the container walls increases. The best microscopic explanation for this pressure increase is that at the smaller volume

   A the individual gas molecules are compressed

   B the gas molecules repel each other more strongly

   C the average velocity of gas molecules hitting the wall is greater

   D the frequency of collisions with gas molecules with the walls is greater

4 A lead bullet is fired into an iron plate, where it deforms and stops. As a result, the temperature of the lead increases by an amount ΔT. For an identical bullet hitting the plate with twice the speed, what is the best estimate of the temperature increase?

   A ΔT **C** $2\,\Delta T$

   B $\sqrt{2}\,\Delta T$ **D** $4\,\Delta T$

5 In winter, in some countries, the water in a swimming pool needs to be heated.

  (a) Estimate the cost of heating the water in a typical swimming pool from 5 °C to a suitable temperature for swimming. You may choose to consider any reasonable size of pool.

    Clearly show any estimated values. The following information will be useful:

    Specific heat capacity of water   4186 J kg^{-1} K^{-1}
    Density of water   1000 kg m^{-3}
    Cost per kW h of electrical energy $0.10

    (i) Estimated values   [4]
    (ii) Calculations   [7]

  (b) An electrical heater for swimming pools has the following information written on its side:

50 Hz	2.3 kW

    (i) Estimate how many days it would take this heater to heat the water in the swimming pool.   [4]
    (ii) Suggest two reasons why this can only be an approximation.   [2]

  (c) Overnight the water in the swimming pool cools. The temperature loss depends on the conditions during the night. Two possible factors affecting the temperature loss are listed below.

    a CLOUDY night or a CLEAR night

    a DAMP night or a DRY night.

    (i) For each of these factors underline which of the extremes given above would cause the smallest temperature loss, and explain your reasoning.   [6]
    (ii) Suggest one other factor that might affect the overnight temperature loss and explain its effect.   [2]

6 This question is about determining the specific latent heat of fusion of ice.

A student determines the specific latent heat of fusion of ice at home. She takes some ice from the freezer, measures its mass and mixes it with a known mass of water in an insulating jug. She stirs until all the ice has melted and measures the final temperature of the mixture. She also measured the temperature in the freezer and the initial temperature of the water.

She records her measurements as follows:

Mass of ice used	m_i	0.12 kg
Initial temperature of ice	T_i	-12 °C
Initial mass of water	m_w	0.40 kg
Initial temperature of water	T_w	22 °C
Final temperature of mixture	T_f	15 °C

The heat capacities of water and ice are c_w = 4.2 kJ kg^{-1} °C^{-1} and c_i = 2.1 kJ kg^{-1} °C^{-1}.

  (a) Set up the appropriate equation, representing energy transfers during the process of coming to thermal equilibrium, that will enable them to solve for the specific latent heat l_i of ice. Insert values into the equation from the data above, **but do not solve the equation**.   [5]

  (b) Explain the physical meaning of each *energy transfer term* in your equation (but not each symbol).   [4]

  (c) State an assumption you have made about the experiment, in setting up your equation in **(a)**.   [1]

  (d) Why should she take the temperature of the mixture *immediately* after all the ice has melted?   [1]

  (e) Explain from the microscopic point of view, in terms of molecular behaviour, why the temperature of the ice does not increase while it is melting.   [4]

Travelling waves

INTRODUCTION – RAYS AND WAVE FRONTS

Light, sound and ripples on the surface of a pond are all examples of wave motion
- they all transfer energy from one place to another.
- they do so without a net motion of the medium through which they travel.
- they all involve oscillations (vibrations) of one sort or another.

A **continuous wave** involves a succession of individual oscillations. A **wave pulse** involves just one oscillation. Two important categories of wave are **transverse** and **longitudinal** (see below). The table gives some examples.

	Example of energy transfer
Water ripples (Transverse)	A floating object gains an 'up and down' motion.
Sound waves (Longitudinal)	The sound received at an ear makes the eardrum vibrate.
Light wave (Transverse)	The back of the eye (the retina) is stimulated when light is received.
Earthquake waves (Both T and L)	Buildings collapse during an earthquake.
Waves along a stretched rope (Transverse)	A 'sideways pulse' will travel down a rope that is held taught between two people.
Compression waves down a spring (Longitudinal)	A compression pulse will travel down a spring that is is held taught between two people.

The following pages analyse some of the properties that are common to all waves.

LONGITUDINAL WAVES

Sound is a longitudinal wave. This is because the oscillations are **parallel** to the direction of energy transfer.

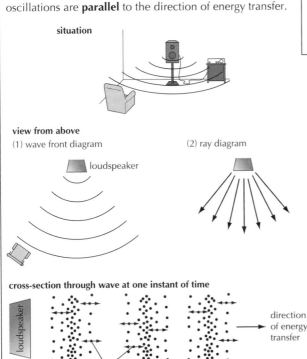

situation

view from above
(1) wave front diagram   (2) ray diagram

loudspeaker

direction of energy transfer

cross-section through wave at one instant of time

motion of air molecules in same direction as energy transfer

wave pattern moves out from loudspeaker

TRANSVERSE WAVES

Suppose a stone is thrown into a pond. Waves spread out as shown below.

situation

(1) wave front diagram

(2) ray diagram

direction of energy flow

cross-section through water

wave pattern moves out from centre

wave pattern at a given instant of time

water surface moves 'up' and 'down'

wave pattern slightly later in time

← centre of pond   edge of pond →

The top of the wave is known as the **crest**, whereas the bottom of the wave is known as the **trough**.

Note that there are several aspects to this wave that can be studied. These aspects are important to all waves.
- The movement of the wave pattern. The **wave fronts** highlight the parts of the wave that are moving together.
- The direction of energy transfer. The **rays** highlight the direction of energy transfer.
- The oscillations of the medium.

It should be noted that the rays are at right angles to the wave fronts in the above diagrams. This is always the case.

This wave is an example of a transverse wave because the oscillations are **at right angles** to the direction of energy transfer.

A point on the wave where everything is 'bunched together' (high pressure) is known as a **compression**. A point where everything is 'far apart' (low pressure) is known as a **rarefaction**.

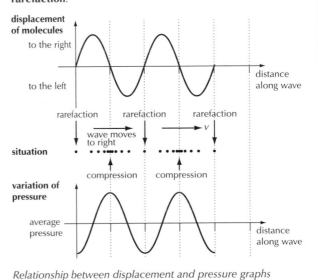

displacement of molecules

to the right

to the left

distance along wave

rarefaction   rarefaction   rarefaction

wave moves to right

situation

compression   compression

variation of pressure

average pressure

distance along wave

Relationship between displacement and pressure graphs

Wave characteristics

DEFINITIONS

There are some useful terms that need to be defined in order to analyse wave motion in more detail. The table below attempts to explain these terms and they are also shown on the graphs.

Because the graphs seem to be identical, you need to look at the axes of the graphs carefully.

- The displacement – time graph represent the oscillations for one point on the wave. All the other points on the wave will oscillate in a similar manner, but they will not start their oscillations at exactly the same time.
- The displacement – position graph represent a 'snap shot' of all the points along the wave at one instant of time. At a later time, the wave will have moved on but it will retain the same shape.
- The graphs can be used to represent longitudinal AND transverse waves because the y-axis records only the value of the displacement. It does NOT specify the direction of this displacement. So, if this displacement were parallel to the direction of the wave energy, the wave would be a longitudinal wave. If this displacement were at right angles to the direction of the wave energy, the wave would be a longitudinal wave.

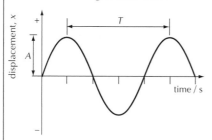 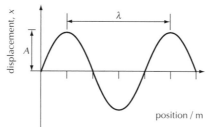

Term	Symbol	Definition
Displacement	x	This measures the change that has taken place as a result of a wave passing a particular point. Zero displacement refers to the mean (or average) position. For mechanical waves the displacement is the distance (in metres) that the particle moves from its undisturbed position.
Amplitude	A	This is the maximum displacement from the mean position. If the wave does not lose any of its energy its amplitude is constant.
Period	T	This is the time taken (in seconds) for one complete oscillation. It is the time taken for one complete wave to pass any given point.
Frequency	f	This is the number of oscillations that take place in one second. The unit used is the Hertz (Hz). A frequency of 50 Hz means that 50 cycles are completed every second.
Wavelength	λ	This is the shortest distance (in metres) along the wave between two points that are **in phase** with one another. 'In phase' means that the two points are moving exactly in step with one another. For example, the distance from one crest to the next crest on a water ripple or the distance from one compression to the next one on a sound wave.
Wave speed	v	This is the speed (in m s⁻¹) at which the wave fronts pass a stationary observer.

The period and the frequency of any wave are inversely related. For example, if the frequency of a wave is 100 Hz, then its period must be exactly $\frac{1}{100}$ of a second.
In symbols,

$$T = \frac{1}{f}$$

WAVE EQUATIONS

There is a very simple relationship that links wave speed, wavelength and frequency. It applies to all waves.

The time taken for one complete oscillation is the period of the wave, T.

In this time, the wave pattern will have moved on by one wavelength, λ.

This means that the speed of the wave must be given by

$$v = \frac{\text{distance}}{\text{time}} = \frac{\lambda}{T}.$$

Since $\frac{1}{T} = f$

$$v = f\lambda$$

In words,

velocity = frequency × wavelength

EXAMPLE

A stone is thrown onto a still water surface and creates a wave. A small floating cork 1.0 m away from the impact point has the following displacement-time graph (time is measured from the instant the stone hits the water):

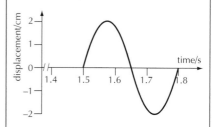

(a) the amplitude of the wave:

2 cm

(b) the velocity of the wave:

$$v = \frac{d}{t} = \frac{1}{1.5} = 0.67 \text{ m s}^{-1}$$

(c) the frequency of the wave:

$$f = \frac{1}{T} = \frac{1}{0.3} = 3.33 \text{ Hz}$$

(d) the wavelength of the wave:

$$\lambda = \frac{v}{f} = \frac{0.666}{3.33} = 0.2 \text{ m}$$

Reflection, refraction and transmission of waves

REFLECTION OF ONE-DIMENSIONAL WAVES

In general, when any wave meets the boundary between two different media it is partially reflected and partially transmitted. The diagrams below represent what happens to one-dimensional transverse wave pulses.

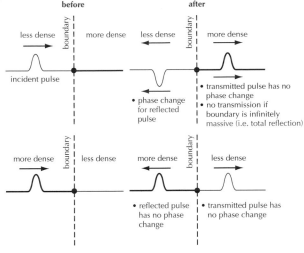

Summary of reflection and transmission for a transverse pulse

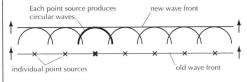

Reflection of a longitudinal pulse

REFLECTION OF TWO-DIMENSIONAL PLANE WAVES

The diagram below shows what happens when plane waves are reflected at a boundary. When working with rays, by convention we always measure the angles between the rays and the **normal**. The normal is a construction line that is drawn at right angles to the surface.

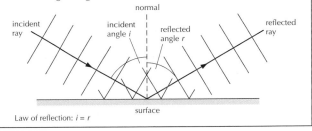

Law of reflection: $i = r$

REFRACTION OF PLANE WAVES

If plane waves are incident at an angle on the boundary between two media, the transmitted wave will change direction – it has been **refracted**. The reason for this change in direction is the change in speed of the wave.

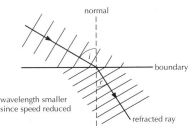

wavelength smaller since speed reduced

Snell's law (an experimental law of refraction) states that the ratio $\dfrac{\sin i}{\sin r}$ = constant, for a given frequency.

HUYGENS' PRINCIPLE

The laws of refraction and reflection can both be derived if we adopt a particular mathematical method called Huygens' principle.

Picture a wave front at one instant in time. If we want to work out where the wave front is at some later time we adopt the following procedure:

- imagine every point on the initial wave front as an individual point source of circular waves.
- construct the waves that each point source would produce.
- imagine these circular waves all add together. This adding is called superposition.
- the new wave front can be constructed as the 'envelope' that contains all these waves.

The diagram below represents this process.

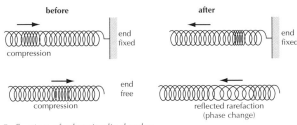

Huygens' construction

Using this construction, we can derive the laws of reflection and refraction.

Huygens' construction and reflection

Huygens' construction and refraction

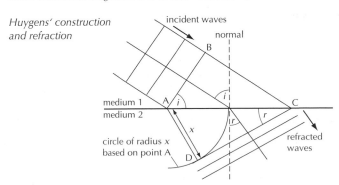

In the time that B takes to travel to C, the wave from the Huygens' point source at A will have travelled the same distance. The new wave front will be CD.
Δ ABC is the mirror image of Δ ADC, so BÂC = DĈA ∴ **$i = r$**

In the time that B takes to travel to C, the wave from the Huygens' point source at A will have travelled a distance, x. using Δ ABC:

$\sin i = \dfrac{BC}{AC}$ using Δ ACD: $\sin r = \dfrac{AD}{AC}$ ∴ $\dfrac{\sin i}{\sin r} = \dfrac{BC}{AC} \div \dfrac{AD}{AC} = \dfrac{BC}{AD} = \dfrac{\text{speed in 1}}{\text{speed in 2}}$

Wave diffraction and interference

DIFFRACTION

When waves pass through apertures they tend to spread out. Waves also spread around obstacles. This wave property is called **diffraction**.

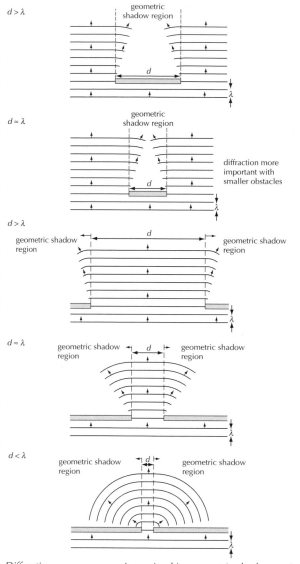

Diffraction – wave energy is received in geometric shadow region
d = width of obstacle/gap

The are some important points to note from these diagrams.

- diffraction becomes relatively more important when the wavelength is large in comparison to the size of the aperture (or the object).
- the wavelength needs to be of the same order of magnitude as the aperture for diffraction to be noticeable.

Huygens' principle gives a good insight into the process of diffraction if the slit is narrow compared to the wavelength.

1 According to Huygens, each point on the wave front acts like a point source.
2 Circular waves start from each point source.
3 A new wave front is formed from all of the waves.
4 Aperture removes all but a single point source.
5 The result is a circular wave.

INTERFERENCE

When two waves of the same type meet, they **interfere** and we can work out the resulting wave using the **principle of superposition**. The overall disturbance at any point and at any time where the waves meet is the vector sum of the disturbances that would have been produced by each of the individual waves. This is shown below.

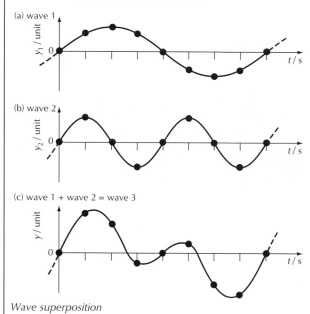

Wave superposition

If the waves have the same amplitude and the same frequency then the interference **at a particular point** can be **constructive** or **destructive**.

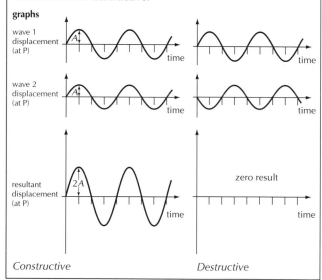

Constructive   *Destructive*

TECHNICAL LANGUAGE

Constructive interference takes place when the two waves are 'in step' with one another – they are said to be **in phase**. There is a zero **phase difference** between them. Destructive interference takes place when the waves are exactly 'out of step' – they are said to be **out of phase**. There are several different ways of saying this. One could say that the phase difference is equal to 'half a cycle' or '180 degrees' or 'π radians'.

For constructive or destructive interference to take place, the sources of the waves must be phase linked or **coherent**.

EXAMPLES OF DIFFRACTION

Diffraction provides the reason why we can hear something even if we can not see it.

If you look at a distant street light at night and then squint your eyes the light spreads sideways – this is as a result of diffraction taking place around your eyelashes! (Needless to say, this explanation is a simplification.)

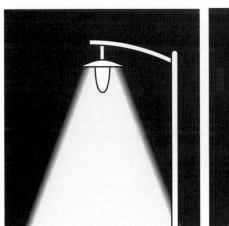

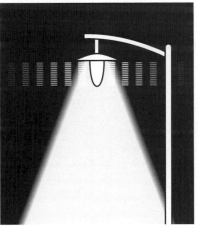

EXAMPLES OF INTERFERENCE

Water waves
A ripple tank can be used to view the interference of water waves. Regions of large-amplitude waves are constructive interference. Regions of still water are destructive interference.

Sound
It is possible to analyse any noise in terms of the component frequencies that exist. A computer can then generate exactly the same frequencies but of different phase. This 'antisound' will interfere with the original sound. An observer in a particular position in space could have the overall noise level reduced if the waves superimposed destructively at that position.

Light
The colours seen on the surface of a soap bubble are a result of constructive and destructive interference of two light rays. One ray is reflected off the outer surface of the bubble whereas the other is reflected off the inner surface.

Doppler effect

DOPPLER EFFECT

The Doppler effect is the name given to the change of frequency of a wave as a result of the movement of the source or the movement of the observer. The details of the effect are shown on page 72, but standard level students do not need all of this. They only need to be able to describe the effect.

The diagram below represents a moving source of sound waves. The source is moving at a speed less than the speed of sound in air.

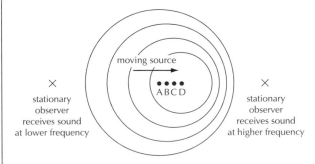

- Sound waves are emitted at a particular frequency from the source.
- The speed of the sound wave in air does not change, but the motion of the source means that the wave fronts are all 'bunched up' ahead of the source.

- This means that the stationary observer receives sound waves of reduced wavelength.
- Reduced wavelength corresponds to an increased frequency of sound.

The overall effect is that the observer will hear sound at a higher frequency than it was emitted by the source. This applies when the source is moving towards the observer. A similar analysis quickly shows that if the source is moving away from the observer, sound of a lower frequency will be received. A change of frequency can also be detected if the source is stationary, but the observer is moving.

- When a police car or ambulance passes you on the road, you can hear the pitch of the sound change from high to low frequency. It is high when it is approaching and low when it is going away.
- Radar detectors can be used to measure the speed of a moving object. They do this by measuring the change in the frequency of the reflected wave.
- For the Doppler effect to be noticeable with light waves, the source (or the observer) needs to be moving at high speed. If a source of light of a particular frequency is moving away from an observer, the observer will receive light of a lower frequency. Since the red part of the spectrum has lower frequency than all the other colours, this is called a **red shift**.
- If the source of light is moving towards the observer, there will be a **blue shift**.

Nature and production of standing waves

STANDING WAVES

A special case of interference occurs when two waves meet that are:
- of the same amplitude
- of the same frequency
- travelling in opposite directions.

In these conditions a **standing wave** will be formed.

The conditions needed to form standing waves seem quite specialised, but standing waves are in fact quite common. They often occur when a wave reflects back from a boundary along the route that it came. Since the reflected wave and the incident wave are of (nearly) equal amplitude, these two waves can interfere and produce a standing wave.

Perhaps the simplest way of picturing a standing wave would be to consider two transverse waves travelling in opposite directions along a stretched rope. The series of diagrams below shows what happens.

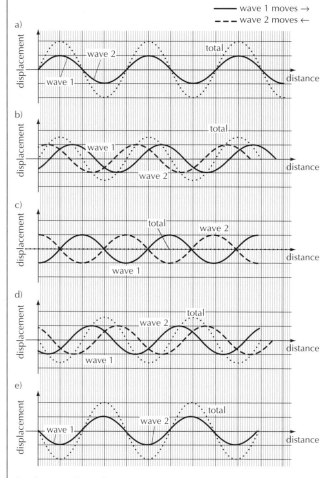

Production of standing waves

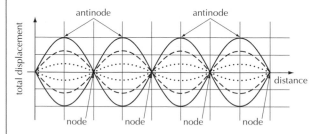

A standing wave – the pattern remains fixed.

There are some points on the rope that are always at rest. These are called the **nodes**. The points where the maximum movement takes place are called **antinodes**. The resulting standing wave is so called because the wave pattern remains fixed in space – it is its amplitude that changes over time. A comparison with a normal (travelling) wave is given below.

	Stationary wave	Normal (travelling) wave
Amplitude	All points on the wave have different amplitudes. The maximum amplitude is 2A at the antinodes. It is zero at the nodes.	All points on the wave have the same amplitude.
Frequency	All points oscillate with the same frequency.	All points oscillate with the same frequency.
Wavelength	This is **twice** the distance from one node (or antinode) to the next node (or antinode).	This is the shortest distance (in metres) along the wave between two points that are in phase with one another.
Phase	All points between one node and the next node are moving in phase.	All points along a wavelength have different phases.
Energy	Energy is not transmitted by the wave, but it does have an energy associated with it.	Energy is transmitted by the wave.

Although the example above involved transverse waves on a rope, a standing wave can also be created using sound or light waves. All musical instruments involve the creation of a standing sound wave inside the instrument. The production of laser light involves a standing light wave. Even electrons in hydrogen atoms can be explained in terms of standing waves.

A standing longitudinal wave can be particularly hard to imagine. The diagram below attempts to represent one example – a standing sound wave.

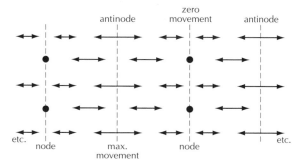

A longitudinal standing wave

Boundary conditions and resonance

RESONANCE

The concept of **resonance** is a very general concept – examples can be found in many areas of Physics. These include the motion of a child on a swing (when the swing is being pushed by an adult) and the tuning of a radio circuit.

In outline, any system that can oscillate can be made to resonate. A system will have its own natural frequency (or frequencies) of oscillation, but it can be forced to oscillate at any frequency if it is driven by another oscillator. Resonance occurs when the driving frequency is equal to the system's own natural frequency. Under these conditions, the amplitude of the oscillations grows and the energy of the system reaches a maximum.

BOUNDARY CONDITIONS

The boundary conditions of the system specify the conditions that must be met at the edges (the boundaries) of the system when standing waves are taking place. Any standing wave that meets these boundary conditions will be a possible resonant mode of the system.

1. Transverse waves on a string.

If the string is fixed at each end, the ends of the string cannot oscillate. Both ends of the string would reflect a travelling wave and thus a standing wave is possible. The only standing waves that fit these boundary conditions are ones that have nodes at each end. The diagrams below show the possible resonant modes.

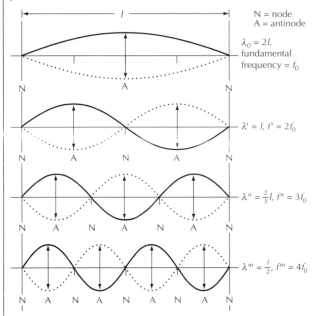

Fundamental and higher resonant modes for a string.

The resonant mode that has the lowest frequency is called the **fundamental** or the **first harmonic**. Higher resonant modes are called **harmonics**. Many musical instruments (e.g. piano, violin, guitar etc) involve similar oscillations of metal 'strings'.

2. Longitudinal sound waves in a pipe.

A longitudinal standing wave can be set-up in the column of air enclosed in a pipe. As in the example above, this results from the reflections that take place at both ends.

As before, the boundary conditions determine the standing waves that can exist in the tube. A closed end must be a displacement node. An open end must be an antinode. Possible standing waves are shown for a pipe open at both ends and a pipe closed at one end.

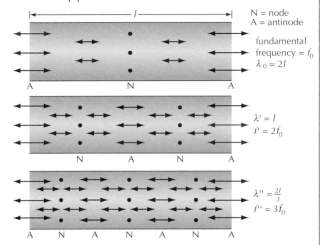

Fundamental and higher resonant modes for a pipe open at both ends.

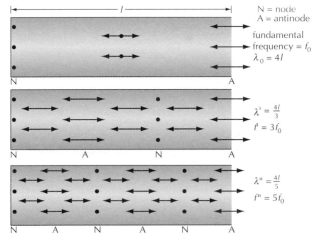

Fundamental and higher resonant modes for a pipe closed at one end.

Musical instruments that involve a standing wave in a column of air include the flute, the trumpet, the recorder, organ pipes etc.

EXAMPLE

An organ pipe (open at one end) is 1.2 m long.
Calculate its fundamental frequency.
The speed of sound is 330 m s^{-1}.

$l = 1.2$ m $\quad \therefore \quad \dfrac{\lambda}{4} = 1.2$ m (fundamental)

$\therefore \quad \lambda = 4.8$ m

$v = f\lambda$

$f = \dfrac{330}{4.8} \simeq 69$ Hz

IB QUESTIONS – WAVES

1 A surfer is out beyond the breaking surf in a deep-water region where the ocean waves are sinusoidal in shape. The crests are 20 m apart and the surfer rises a vertical distance of 4.0 m from wave **trough** to **crest**, in a time of 2.0 s. What is the speed of the waves?

A 1.0 m s^{-1} **B** 2.0 m s^{-1} **C** 5.0 m s^{-1} **D** 10.0 m s^{-1}

2 Radio waves of wavelength 30 cm have a frequency of about

A 10 MHz **B** 90 MHz **C** 1000 MHz **D** 9000 MHz

3 Light passes through three different media, being refracted at each interface as shown.

Which of the options below correctly indicates how the speed of light compares in the three media?

A $v_1 > v_2 > v_3$ **C** $v_3 > v_1 > v_2$

B $v_3 > v_2 > v_1$ **D** $v_2 > v_1 > v_3$

4 Microwaves of wavelength 4.0 cm are emitted normally towards a reflecting surface and they are reflected back. A detector, which measures the net microwave intensity, moves along the line joining the emitter and reflector as shown below.

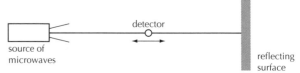

The distance moved by the detector between one point of minimum intensity and **the next** minimum point will be

A 0.5 cm **B** 1.0 cm **C** 2.0 cm **D** 4.0 cm

5 In order to measure the speed of sound in water a loudspeaker and microphone are set up to float in the middle of a swimming pool as shown in **Figure A** below. The microphone output is recorded directly by an oscilloscope and the recording electronics. The water depth is measured as 0.85 m.

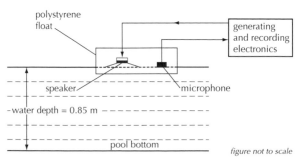

Figure A

The experiment proceeds as follows: At some time, call it $t = 0$, a pulse of sound is generated in the speaker. **At the same time** the recording equipment is triggered to start recording. The burst of sound is of 1.0 ms duration and travels to the bottom of the pool and is reflected, back to the top, where it is detected by the microphone.

A typical recorded signal is shown in **Figure B** below.

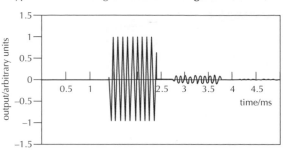

Figure B

(a) Describe the features of the whole recorded signal. (At least **three** different points should be noted.) [3]

(b) From the recorded data, determine the speed of sound in the water. [4]

(c) Why is it best to do this experiment in the middle of the pool and not near the sides? [1]

(d) What is the frequency of the sound wave in the 1.0 m s^{-1} sound pulse? [2]

(e) What is the wavelength of the sound in water at the frequency used? [1]

(f) If the frequency of the sound wave in the pulse was changed, what effect would this have on the measured speed of sound? [1]

(g) A more precise value for the speed of sound in water could be obtained by using a number of different water depths, particularly the greater depths in the diving pool. Why would greater depths give more precise values? [2]

(h) Consider whether a similar pulse timing technique could be used to measure the speed of light by replacing the speaker with a light source and the microphone with a light detector. Suggest at least **two** difficulties that would be experienced in trying to carry out such an experiment. [3]

6 The note played by a violin depends on a number of factors. A violinist can change the note by adjusting its tension or by adjusting its length.

A particular violin string has a length of 0.400 m. It produces a note of fundamental frequency 440 Hz, together with higher harmonics.

(a) Sketch the pattern of vibration for the fundamental, and one of the harmonic modes. [2]

(b) Calculate the wavelength of the fundamental on the string. [1]

(c) Calculate the frequency of the harmonic that you have drawn. [1]

(d) The violinist now wishes to play a note of fundamental frequency 524 Hz on the same string. This is done by using a finger to shorten the effective vibrating length of the string. Determine where on the string the finger must be placed in order to produce the new note. [3]

Electric charge

CONSERVATION OF CHARGE

Two types of charge exist – positive and negative. Equal amounts of positive and negative charge cancel each other. Matter that contains no charge, or matter that contains equal amounts of positive and negative charge, is said to be electrically **neutral**.

Charges are known to exist because of the forces that exist between all charges, called the **electrostatic force**: like charges repel, unlike charges attract.

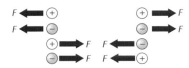

Like charges repel, unlike charges attract.

A very important experimental observation is that charge is always conserved.

Charged objects can be created by friction. In this process electrons are physically moved from one object to another – in order for the charge to remain on the object, it normally needs to be an insulator.

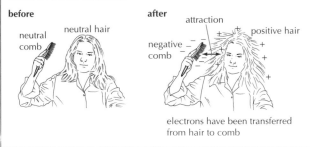

electrons have been transferred from hair to comb

UNITS

The smallest amount of negative charge available is the charge on an electron; the smallest amount of positive charge is the charge on a proton. In everyday situations this unit would be far too small so we use the **coulomb, C**. One coulomb of negative charge would be the charge carried by a total of 6¼ million, million, million electrons (6.25×10^{18}).

INDUCTION

Apart from friction, there are two main processes by which an object gains a charge: charging by contact and charging by **induction**. Induction is the name given to a process of charge separation.

In the diagrams below a positive rod is used to give an electroscope an overall negative charge.

CONDUCTORS AND INSULATORS

A material that allows the flow of charge through it is called an electrical **conductor**. If charge cannot flow through a material it is called an electrical **insulator**. In conductors the flow of charge is always as a result of the flow of electrons from atom to atom.

Electrical conductors	Electrical insulators
all metals e.g. copper aluminium brass graphite	plastics e.g. polythene nylon acetate rubber dry wood glass ceramics

THE GOLD LEAF ELECTROSCOPE

A simple instrument used to demonstrate the presence of a charged object is the **gold leaf electroscope**. The central metal cap, rod and leaf system is insulated from a metal case. If the leaf is given a charge, the leaf lifts away from the rod – it diverges.

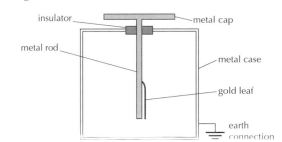

The gold leaf electroscope

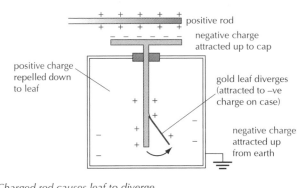

Charged rod causes leaf to diverge

1 neutral electroscope

2 charged rod causes leaf to diverge

3 Touching the cap causes the leaf to go back down (negative charge flows from earth through the finger).

4 Removing the finger has no effect on the charge distribution.

5 When the rod is removed, the electroscope has gained an overall negative charge on the cap–rod–leaf system.

Electric force and electric potential

COULOMB'S LAW

The diagram shows the force between two point charges that are far away from the influence of any other charges.

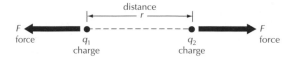

The directions of the forces are along the line joining the charges. If they are like charges, the forces are away from each other – they repel. If they are unlike charges, the forces are towards each other – they attract.

Each charge must feel a force of the same size as the force on the other one

Experimentally, the force is proportional to the size of both charges and inversely proportional to the square of the distance between the charges.

$$F = \frac{k\, q_1\, q_2}{r^2}$$

This is known as Coulomb's law and the constant k is called the Coulomb constant. In fact, the law is often quoted in a slightly different form using a different constant for the medium called the permittivity, ε.

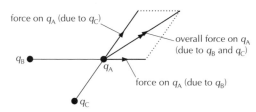

If there are two or more charges near another charge, the overall force can be worked out using vector addition.

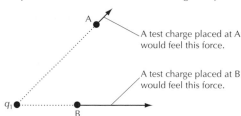

Vector addition of electrostatic forces

ELECTRIC FIELDS – DEFINITION

A charge, or combination of charges, is said to produce an **electric field** around it. If we place a **test charge** at any point in the field, the value of the force that it feels at any point will depend on the value of the test charge only.

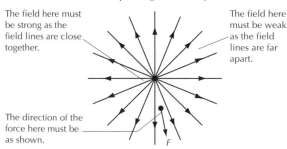

A test charge would feel a different force at different points around a charge q_1.

In practical situations, the test charge needs to be small so that it doesn't disturb the charge or charges that are being considered.

The definition of electric field, E, is

$$E = \frac{F}{q_2} = \text{force per positive unit test charge.}$$

Coulomb's law can be used to relate the electric field around a point charge to the charge producing the field.

$$E = \frac{q_1}{4\pi\varepsilon_0 r^2}$$

When using these equations you have to be very careful
- not to muddle up the charge producing the field and the charge sitting in the field (and thus feeling a force).
- not to use the mathematical equation for the field around a point charge for other situations (e.g. parallel plates).

REPRESENTATION OF ELECTRIC FIELDS

This is done using **field lines**.

At any point in a field
- the direction of field is represented by the direction of the field lines closest to that point.
- the magnitude of the field is represented by the number of field lines passing near that point.

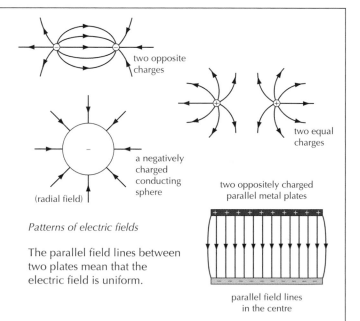

Field around a positive point charge

The resultant electric field at any position due to a collection of point charges is shown to the right.

Patterns of electric fields

The parallel field lines between two plates mean that the electric field is uniform.

Electric potential energy and electric potential difference

ENERGY DIFFERENCE IN AN ELECTRIC FIELD

When placed in an electric field, a charge feels a force. This means that if it moves around in an electric field work will be done. As a result, the charge will either gain or lose electrical potential energy. Electrical potential energy is the energy that a charge has as a result of its position in an electrical field. This is the same idea as a mass in a gravitational field. If we lift a mass up, its gravitational potential energy increases. If the mass falls, its gravitational potential energy decreases. In the example below a positive charge is moved from position A to position B. This results in an increase in electrical potential energy. Since the field is uniform, the force is constant. This makes it is very easy to calculate the work done.

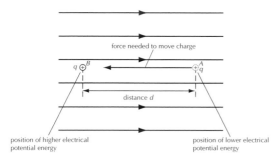

Charge moving in an electric field

Change in electrical potential energy = force × distance

$$= E\,q \times d$$

In the example above the electrical potential energy at B is greater than the electrical potential energy at A. We would have to put in this amount of work to push the charge from A to B. If we let go of the charge at B it would be pushed by the electric field. This push would accelerate it so that the loss in electrical potential energy would be the same as the gain in kinetic energy.

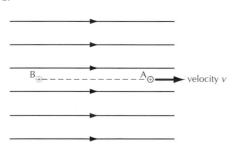

A positive charge released at B will be accelerated as it travels to point A.

gain in kinetic energy = loss in electrical potential energy

$$\tfrac{1}{2}mv^2 = Eqd$$

$$mv^2 = 2Eqd$$

$$\therefore v = \sqrt{\frac{2Eqd}{m}}$$

ELECTRICAL POTENTIAL DIFFERENCE.

In the example left, the actual energy difference between A and B depended on the charge that was moved. If we doubled the charge we would double the energy difference. The quantity that remains fixed between A and B is the energy difference **per unit charge**. This is called the **potential difference**, or **p.d.**, between the points.

$$\frac{\text{Potential difference}}{\text{between two points}} = \frac{\text{energy difference}}{\text{per unit charge moved}}$$

$$= \frac{\text{energy difference}}{\text{charge}}$$

The basic unit for potential difference is the joule/coulomb, $J\,C^{-1}$. A very important point to note is that for a given electric field, the potential difference between any two points is a single fixed scalar quantity. The work done between these two points does not depend on the path taken by the test charge. A technical way of saying this is 'the electric field is **conservative**'.

UNITS

From its definition, the units of potential difference (p.d.) are $J\,C^{-1}$. This is given a new name, the volt, V. Thus:

1 volt = 1 J C^{-1}

Many people use this unit to give potential difference a new name 'voltage'. Voltage and potential difference are different words for the same thing. Potential difference is probably the better name to use as it reminds you that it is measuring the difference between two points.

When working at the atomic scale, the joule is far too big to use for a unit for energy. The everyday unit used by physicists for this situation is the electronvolt. As could be guessed from its name, the electronvolt is simply the energy that would be gained by an electron moving through a potential difference of 1 volt.

Since energy gained = p.d. × charge

$$1 \text{ electronvolt} = 1 \text{ volt} \times 1.6 \times 10^{-19} \text{ C}$$

$$= 1.6 \times 10^{-19} \text{ J}$$

The normal SI prefixes also apply so one can measure energies in kiloelectronvolts (keV) or megaelectronvolts (MeV). The latter unit is very common in particle physics.

EXAMPLE

Calculate the speed of an electron accelerated in a vacuum by a p.d. of 1000V.

$$\text{KE of electron} = V \times e$$

$$= 1000 \times 1.6 \times 10^{-19}$$

$$= 1.6 \times 10^{-16} \text{ J}$$

$$\tfrac{1}{2}mv^2 = 1.6 \times 10^{-16} \text{ J}$$

$$v^2 = \frac{2 \times 1.6 \times 10^{-16}}{9.11 \times 10^{-31}}$$

$$v = 5.9 \times 10^5 \text{ ms}^{-1}$$

Electric current

ELECTRICAL CONDUCTION IN A METAL

Whenever charges move we say that a **current** is flowing. A current is the name for moving charges and the path that they follow is called the **circuit**. Without a complete circuit, a current cannot be maintained for any length of time.

Current flows THROUGH an object when there is a potential difference ACROSS the object. A battery (or power supply) is the device that creates the potential difference.

By convention, currents are always represented as the flow of positive charge. Thus **conventional current**, as it is known, flows from positive to negative. Although currents can flow in solids, liquids and gases, in most everyday electrical circuits the currents flow through wires. In this case the things that actually move are the negative electrons – the **conduction electrons**. The direction in which they move is opposite to the direction of the representation of conventional current. As they move the interactions between the conduction electrons and the lattice ions means that work needs to be done. Therefore, when a current flows, the metal heats up. The speed of the electrons due to the current is called their **drift velocity**.

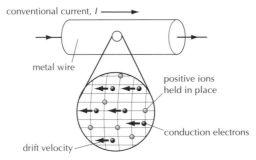

Electrical conduction in a metal

CURRENT

Current is defined as the **rate of flow of electrical charge**. It is always given the symbol, *I*. Mathematically the definition for current is expressed as follows:

$$\text{Current} = \frac{\text{charge flowed}}{\text{time taken}}$$

$$I = \frac{Q}{t} \text{ or (in calculus notation) } I = \frac{dQ}{dt}$$

$$\textbf{1 amp} = \frac{\textbf{1 coulomb}}{\textbf{1 second}}$$

If a current flows in just one direction it is known as a **direct current**. A current that constantly changes direction (first one way then the other) is known as an **alternating current** or a.c.

RESISTANCE

Resistance is the mathematical ratio between potential difference and current. If something has a high resistance, it means that you would need a large potential difference across it in order to get a current to flow.

$$\text{Resistance} = \frac{\text{potential difference}}{\text{current}}$$

In symbols, $R = \frac{V}{I}$

We define a new unit, the ohm, Ω, to be equal to one volt per amp.

$$1 \text{ ohm} = 1 \text{ V A}^{-1}$$

OHM'S LAW – OHMIC AND NON-OHMIC BEHAVIOUR

The graphs below show how the current varies with potential difference for some typical devices.

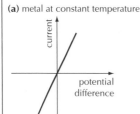
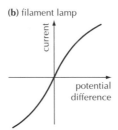
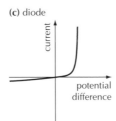

(a) metal at constant temperature **(b)** filament lamp **(c)** diode

If current and potential difference are proportional (like the metal at constant temperature) the device is said to be **ohmic**. Devices where current and potential difference are not proportional (like the filament lamp or the diode) are said to be **non-ohmic**.

Ohm's law states that the current flowing through a piece of metal is proportional to the potential difference across it providing the temperature remains constant.

In symbols,

$$V \propto I \text{ [if temperature is constant]}$$

A device with constant resistance (in other words an ohmic device) is called a **resistor**.

POWER DISSIPATION

Since potential difference $= \dfrac{\text{energy difference}}{\text{charge flowed}}$

And current $= \dfrac{\text{charge flowed}}{\text{time taken}}$

This means that potential difference $\times$ current

$$= \frac{(\text{energy difference})}{(\text{charge flowed})} \times \frac{(\text{charge flowed})}{(\text{time taken})}$$

$$= \frac{\text{energy difference}}{\text{time}}$$

This energy difference per time is the power dissipated by the resistor. All this energy is going into heating up the resistor. In symbols:

$$P = V \times I$$

Sometimes it is more useful to use this equation in a slightly different form, e.g.

$$P = V \times I \text{ but } V = I \times R \text{ so}$$

$$P = (I \times R) \times I$$

$$P = I^2 R$$

Similarly $P = \dfrac{V^2}{R}$

EXAMPLE

A 1.2 kW electric kettle is plugged into the 250 V mains supply. Calculate

(i) the current drawn
(ii) its resistance

(i) $I = \dfrac{1200}{250} = 4.8 \text{ A}$

(ii) $R = \dfrac{250}{4.8} = 52 \ \Omega$

Electric circuits

CIRCUITS

An electrical circuit can contain many different devices or **components**. The mathematical relationship $V = IR$ can be applied to any component or groups of components in a circuit.

When analysing a circuit it is important to look at the circuit as a whole. The power supply is the device that is providing the energy, but it is the whole circuit that determines what current flows through the circuit.

RESISTORS IN SERIES

A **series circuit** has components connected one after another in a continuous chain. The current must be the same everywhere in the circuit since charge is conserved. The total potential difference is shared amongst the components.

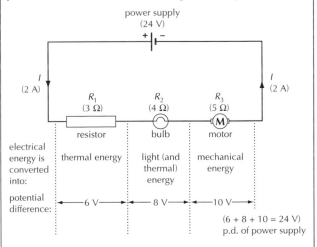

Example of a series circuit

We can work out what share they take by looking at each component in turn, e.g.

The potential difference across the resistor = $I \times R_1$

The potential difference across the bulb = $I \times R_2$

$R_{total} = R_1 + R_2 + R_3$

This always applies to a series circuit. Note that $V = IR$ correctly calculates the potential difference across each individual component as well as calculating it across the total.

RESISTORS IN PARALLEL

A **parallel circuit** branches and allows the charges more than one possible route around the circuit.

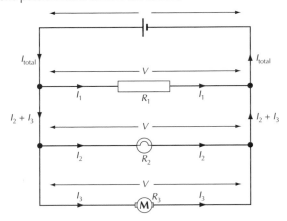

Example of a parallel circuit

Since the power supply fixes the potential difference, each component has the same potential difference across it. The total current is just the addition of the currents in each branch.

$$I_{total} = I_1 + I_2 + I_3$$

$$= \frac{V}{R_1} + \frac{V}{R_2} + \frac{V}{R_3}$$

$$\frac{1}{R_{total}} = \frac{1}{R_1} + \frac{1}{R_2} + \frac{1}{R_3}$$

ELECTRICAL METERS

A current-measuring meter is called an **ammeter**. It should be connected in series at the point where the current needs to be measured. A perfect ammeter would have zero resistance.

A meter that measures potential difference is called a **voltmeter**. It should be placed in parallel with the component or components being considered. A perfect voltmeter has infinite resistance.

ELECTROMOTIVE FORCE AND INTERNAL RESISTANCE

When a 6V battery is connected in a circuit some energy will be used up inside the battery itself. In other words, the battery has some **internal resistance**. The TOTAL energy difference per unit charge around the circuit is still 6 volts, but some of this energy is used up inside the battery. The energy difference per unit charge from one terminal of the battery to the other is less than the total made available by the chemical reaction in the battery.

For historical reasons, the TOTAL energy difference per unit charge around a circuit is called the **electromotive force (e.m.f.)**. However, remember that it is not a force (measured in newtons) but an energy difference per charge (measured in volts).

In practical terms, e.m.f. is exactly the same as potential difference if no current flows.

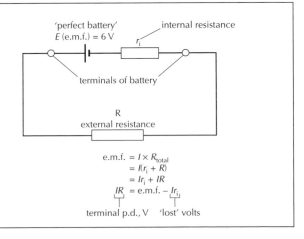

Magnets and magnetic fields

MAGNETIC FIELD LINES

There are many similarities between the magnetic force and the electrostatic force. In fact, both forces have been shown to be two aspects of one force – the electromagnetic interaction (see page 87). It is, however, much easier to consider them as completely separate forces to avoid confusion.

Page 40 introduced the idea of electric fields. A similar concept is used for magnetic fields. A table of the comparisons between these two fields is shown below.

	Electric field	**Magnetic field**
Symbol	E	B
Caused by …	Charges	Magnets (or electric currents)
Affects …	Charges	Magnets (or electric currents)
Two types of …	Charge: positive and negative	Pole: North and South
Simple force rule:	Like charges repel, Unlike charges attract	Like poles repel, Unlike poles attract

In order to help visualise magnetic field we, once again, use the concept of field lines. This time the field lines are lines of magnetic field – also called **flux** lines. If a 'test' magnetic North pole is place in a magnetic field, it will feel a force.

- The direction of the force is shown by the direction of the field lines.
- The strength of the force is shown by how close the lines are to one another.

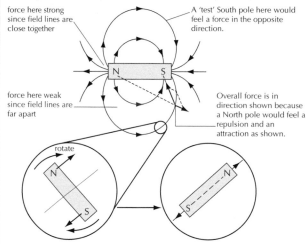

force here strong since field lines are close together

A 'test' South pole here would feel a force in the opposite direction.

force here weak since field lines are far apart

Overall force is in direction shown because a North pole would feel a repulsion and an attraction as shown.

rotate

A small magnet placed in the field would rotate until lined up with the field lines. This is how a compass works. Small pieces of iron (iron filings) will also line up with the field lines – they will be induced to become little magnets.

Field pattern of an isolated bar magnet

Despite all the similarities between electric fields and magnetic fields, it should be remembered that they are very different. For example:
- A magnet does not feel a force when placed in an electric field.
- A positive charge does not feel a force when placed stationary in a magnetic field.
- Isolated charges exist whereas isolated poles do not.

The Earth itself has a magnetic field. It turns out to be similar to that of a bar magnet with a magnetic South pole near the geographic North Pole as shown above right.

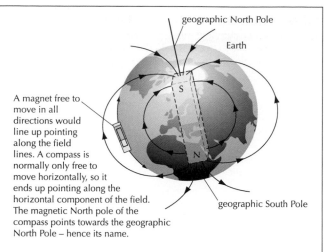

geographic North Pole

Earth

A magnet free to move in all directions would line up pointing along the field lines. A compass is normally only free to move horizontally, so it ends up pointing along the horizontal component of the field. The magnetic North pole of the compass points towards the geographic North Pole – hence its name.

geographic South Pole

An electric current can also cause a magnetic field. The mathematical value of the magnetic fields produced in this way is given on page 46. The field patterns due to different currents can be seen in the diagrams below.

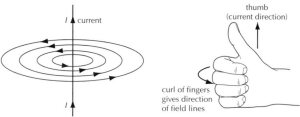

thumb (current direction)

I current

curl of fingers gives direction of field lines

The field lines are circular around the current.

The direction of the field lines can be remembered with the right hand grip rule. If the thumb of the right hand is arranged to point along the direction of a current, the way the fingers of the right hand naturally curl will give the direction of the field lines.

Field pattern of a straight wire carrying current

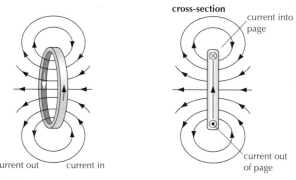

cross-section

current into page

current out   current in

current out of page

Field pattern of a flat circular coil

A long current-carrying coil is called a solenoid.

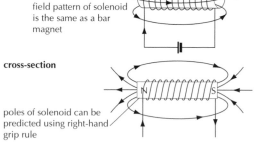

field pattern of solenoid is the same as a bar magnet

cross-section

poles of solenoid can be predicted using right-hand grip rule

Field pattern for a solenoid

Magnetic forces

MAGNETIC FORCE ON A CURRENT

When a current-carrying wire is placed in a magnetic field the magnetic interaction between the two results in a force. This is known as the **motor effect**. The direction of this force is at right angles to the plane that contains the field and the current as shown below.

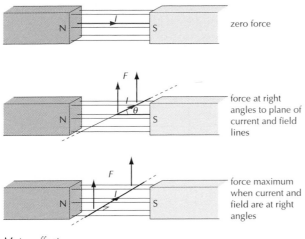

zero force

force at right angles to plane of current and field lines

force maximum when current and field are at right angles

Motor effect

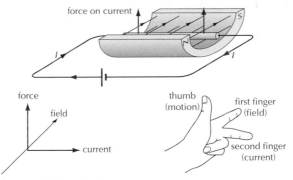

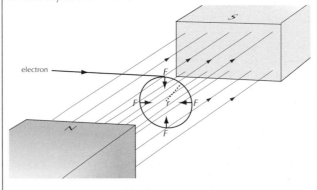

Fleming's left-hand rule

Experiments show that the force is proportional to:
- the magnitude of the magnetic field, B.
- the magnitude of the current, I.
- the length of the current, l, that is in the magnetic field.
- the sine of the angle, θ, between the field and current.

The magnetic field strength, B is defined as follows:

$$F = BIl \sin \theta \quad \text{or}$$

$$B = \frac{F}{IL \sin \theta}$$

A new unit, the tesla, is introduced. 1 T is defined to be equal to 1 N A^{-1} m^{-1}. Another possible unit for magnetic field strength is Wb m^{-2}. Another possible term is magnetic flux density.

D.C. MOTOR

If a current flows around a coil of wire that is placed in a magnetic field, it will begin to rotate, but this rotation will not continue.

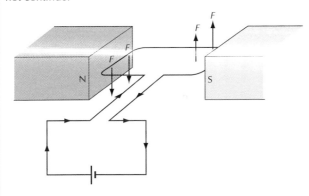

In order to achieve continuous rotation, the current needs to change direction every half revolution. This is achieved with the addition of the **commutator** and the **brushes**. The arrangement is shown below.

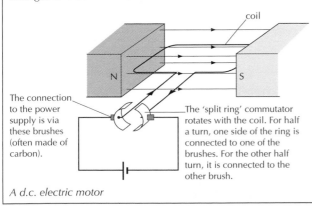

coil

The connection to the power supply is via these brushes (often made of carbon).

The 'split ring' commutator rotates with the coil. For half a turn, one side of the ring is connected to one of the brushes. For the other half turn, it is connected to the other brush.

A d.c. electric motor

MAGNETIC FORCE ON A MOVING CHARGE

A single charge moving through a magnetic field also feels a force in exactly the same way that a current feels a force.

In this case the force on a moving charge is proportional to:
- the magnitude of the magnetic field, B.
- the magnitude of the charge, q.
- the velocity of the charge, v.
- the sine of the angle, θ, between the velocity of the charge and the field.

We can use these relationships to give an alternative definition of the magnetic field strength, B. This definition is exactly equivalent to the previous definition.

$$F = Bqv \sin \theta \quad \text{or} \quad B = \frac{F}{qv \sin \theta}$$

Since the force on a moving charge is always at right angles to the velocity of the charge the resultant motion can be circular. An example of this would be when an electron enters a region where the magnetic field is at right angles to its velocity as shown below.

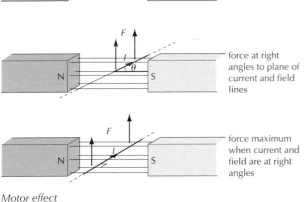

An electron moving at right angles to a magnetic field

The magnetic field due to currents

STRAIGHT WIRE

The field pattern around a long straight wire shows that as one moves away from the wire, the strength of the field gets weaker. Experimentally the field is proportional to
- the value of the current, I.
- the inverse of the distance away from the wire, r. If the distance away is doubled, the magnetic field will halve.

The field also depends on the medium around the wire. These factors are summarised in the equation:

$$B = \frac{\mu I}{2\pi r}$$

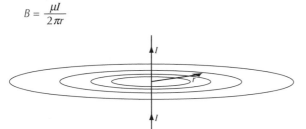

Magnetic field of a straight current

The constant μ is called the permeability and changes if the medium around the wire changes. Most of the time we consider the field around a wire when there is nothing there – so we use the value for the permeability of a vacuum, μ_0. There is almost no difference between the permeability of air and the permeability of a vacuum. There are many possible units for this constant, but it is common to use $N\,A^{-2}$ or $T\,m\,A^{-1}$.

Permeability and permittivity are related constants. In other words, if you know one constant you can calculate the other. In the SI system of units, the permeability of a vacuum is defined to have a value of exactly $4\pi \times 10^{-7}\,N\,A^{-2}$. See the definition of the ampere (right) for more detail.

MAGNETIC FIELD IN A SOLENOID

The magnetic field of a solenoid is very similar to the magnetic field of a bar magnet. As shown by the parallel field lines, the magnetic field inside the solenoid is constant. It might seem surprising that the field does not vary at all inside the solenoid, but this can be experimentally verified near the centre of a long solenoid. It does tend to decrease near the ends of the solenoid as shown in the graph below.

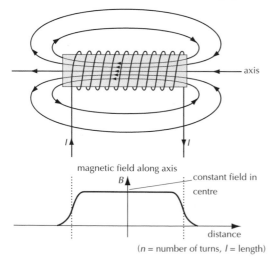

Variation of magnetic field in a solenoid

TWO PARALLEL WIRES – DEFINITION OF THE AMPERE

Two parallel current-carrying wires provide a good example of the concepts of magnetic field and magnetic force. Because there is a current flowing down the wire, each wire is producing a magnetic field. The other wire is in this field so it feels a force. The forces on the wires are an example of a Newton's third law pair of forces.

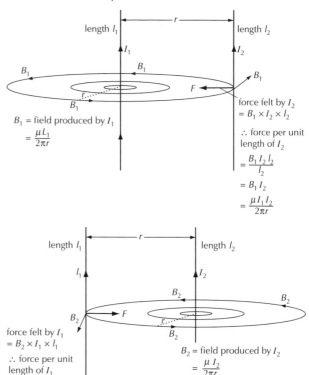

Magnitude of force per unit length on either wire $= \dfrac{\mu I_1 I_2}{2\pi r}$

This equation is experimentally used to define the ampere. The coulomb is then defined to be one ampere second. If we imagine two infinitely long wires carrying a current of one amp separated by a distance of one metre, the equation would predict the force per unit length to be $2 \times 10^{-7}\,N$. Although it is not possible to have infinitely long wires, an experimental set-up can be arranged with very long wires indeed. This allows the forces to be measured and ammeters to be properly calibrated.

The mathematical equation for this constant field at the centre of a long solenoid is

$$B = \mu \left(\frac{n}{l}\right) I$$

Thus the field only depends on:
- the current, I
- the number of turns per unit length, $\frac{n}{l}$
- the nature of the solenoid core, μ

It is independent of the cross-sectional area of the solenoid.

IB QUESTIONS – ELECTRICITY AND MAGNETISM

1 A circuit consists of a battery and three resistors as shown below. The currents at different parts of the circuit are labelled. Which of the following gives a correct relationship between currents?

A $I_2 = I_3$ **B** $I_1 = I_2$ **C** $4I_3 = I_2$ **D** $4I_2 = I_3$

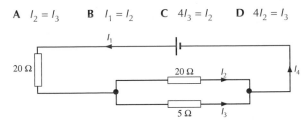

2 The identical lamps in the circuit below each have a resistance R at their rated voltage of 6 V. The lamps are to be run in parallel with each other using a 12 V source and a series resistor Z in the circuit as shown.

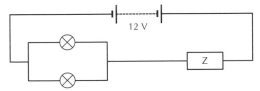

If the lamps are to operate at their rated voltage, what must be the resistance of Z?

A $\dfrac{R}{2}$ **B** R **C** $2R$ **D** zero

3 The element of an electric heater has a resistance R when in operation. What is the resistance of the element of a second heater which has twice the power output at the same voltage?

A $\dfrac{R}{2}$ **B** R **C** $2R$ **D** $4R$

4 In the following circuit, the voltmeter has an internal resistance of 100 kΩ.

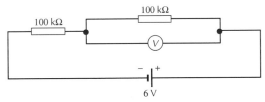

The reading on the voltmeter will be

A 1 V   **B** 2 V   **C** 3 V   **D** 6 V

5 Two uncharged insulated metal spheres, X and Y, are in contact with each other. A positively charged rod R is brought close to X as shown in **Figure I**. Y is now moved away from X (**Figure II**).

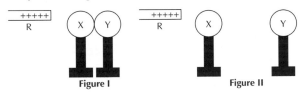

What are the final charge states of X and Y?

	X	Y
A	neutral	neutral
B	positive	neutral
C	neutral	positive
D	negative	positive

6 This question involves physical reasoning and calculations for electric circuits.

Light bulbs are marked with the rating 10 V; 3 W. Suppose you connect three of the bulbs in series with a switch and a 30 V battery as shown in **Figure 1** below. Switch **S** is initially open.

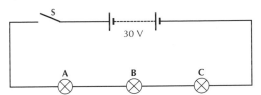

Figure 1

(a) A student tells you that after switch **S** is closed, bulb **C** will light up first, because electrons from the negative terminal of the battery will reach it first, and then go on to light bulbs **B** and **A** in succession. Is this prediction and reasoning correct? How would you reply? [2]

(b) State how the brightness of the three bulbs in the circuit will compare with each other. [1]

(c) The student now connects a fourth bulb **D** across bulb **B** as shown in **Figure 2** below.

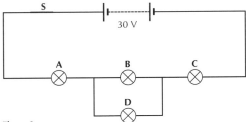

Figure 2

When she connects **D**, what will happen to the brightnesses of bulbs **A, B** and **C**? Explain your reasoning. [3]

(d) Assuming that the resistance of the bulbs remains constant, calculate the power output of bulb **B** in the modified circuit in **Figure 2**. [3]

7 The diagram shows two horizontal parallel wires A and B each 250 mm long. Wire A is rigidly fixed a distance 5 mm vertically above wire B. Wire B lies on a surface with light flexible connecting wires attached to it.

A fixed current of 8 A flows in wire A. The current in wire B is gradually increased until B just starts to lift off the surface, and is then kept constant.

(a) For wire B to lift, must the currents in A and B be in the same or opposite directions? [1]

(b) If the mass of wire B is 0.4 g, determine the minimum current required to lift it off the surface. [5]

(c) After wire B has just lifted off the surface, it *accelerates* towards wire A, even though the current is not further increased. Explain why it accelerates, rather than staying just above the surface or rising at constant speed, and state whether the acceleration is constant or not. [2]

Atomic structure 1

INTRODUCTION

All matter that surrounds us, living or otherwise, is made up of different combinations of atoms. There are only a hundred, or so, different types of atoms present in nature. Atoms of a single type form an element. Each of these elements has a name and a chemical symbol e.g. hydrogen, the simplest of all the elements, has the chemical symbol H. Oxygen has the chemical symbol O. The combination of two hydrogen atoms with one oxygen atom is called a water molecule – H_2O. The full list of elements is shown in a periodic table. Atoms consist of a combination of three things: protons, neutrons and electrons.

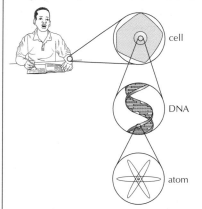

In the basic atomic model, we are made up of protons, neutrons, and electrons – nothing more.

ATOMIC MODEL

The basic atomic model, known as the nuclear model, was developed during the last century and describes a very small central nucleus surrounded by electrons arranged in different energy levels. The nucleus itself contains protons and neutrons (collectively called **nucleons**). All of the positive charge and almost all the mass of the atom is in the nucleus. The electrons provide only a tiny bit of the mass but all of the negative charge. Overall an atom is neutral. The vast majority of the volume is nothing at all – a vacuum. The nuclear model of the atom seems so strange that there must be good evidence to support it.

	Protons	Neutrons	Electrons
Relative mass	1	1	Negligible
Charge	+ 1	Neutral	– 1

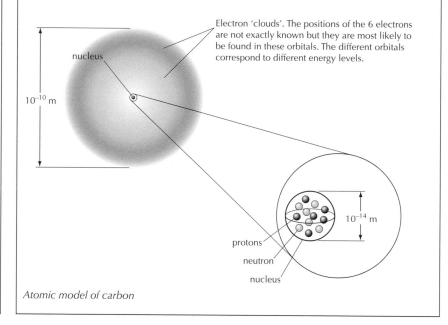

Electron 'clouds'. The positions of the 6 electrons are not exactly known but they are most likely to be found in these orbitals. The different orbitals correspond to different energy levels.

Atomic model of carbon

EVIDENCE

One of the most convincing pieces of evidence for the nuclear model of the atom comes from the Geiger-Marsden experiment. Positive alpha particles were "fired" at a thin gold leaf. The relative size and velocity of the alpha particles meant that most of them were expected to travel straight through the gold leaf. The idea behind this experiment was to see if there was any detectable structure within the gold atoms. The amazing discovery was that some of the alpha particles were deflected through huge angles. The mathematics of the experiment showed that numbers being deflected at any given angle agreed with an inverse square law of repulsion from the nucleus. Evidence for electron energy levels comes from emission and absorption spectra. The existence of isotopes provides evidence for neutrons.

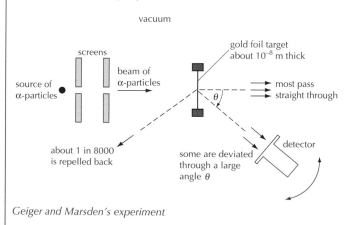

Geiger and Marsden's experiment

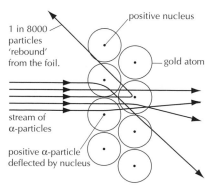

NB not to scale. Only a minute percentage of α-particles are scattered or rebound.

Atomic explanation of Geiger and Marsden's experiment

Atomic structure 2 – Emission and absorption spectra

EMISSION SPECTRA AND ABSORPTION SPECTRA

When an element is given enough energy it emits light. This light can be analysed by splitting it into its various colours (or frequencies) using a prism or a diffraction grating. If all possible frequencies of light were present, this would be called a **continuous spectrum** (see optics section – option H). The light an element emits, its **emission spectrum**, is not continuous, but contains only a few characteristic colours. The frequencies emitted are particular to the element in question. For example, the yellow-orange light from a street lamp is often a sign that the element sodium is present in the lamp. Exactly the same particular frequencies are **absent** if a continuous spectrum of light is shone through an element when it is in gaseous form. This is called an **absorption** spectrum.

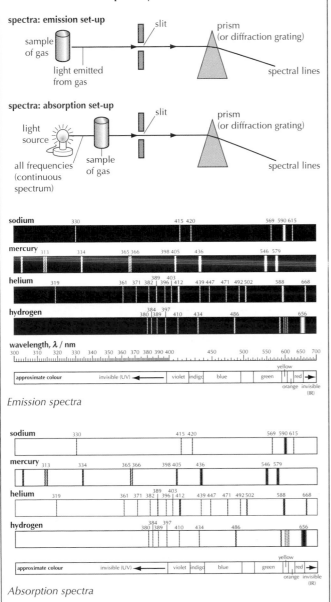

Emission spectra

Absorption spectra

EXPLANATION OF ATOMIC SPECTRA

In an atom, electrons are bound to the nucleus. This means that they cannot "escape" without the input of energy. If enough energy is put in, an electron can leave the atom. If this happens, the atom is now positive overall and is said to be ionised. Electrons can only occupy given energy levels – the energy of the electron is said to be **quantized**. These energy levels are fixed for particular elements and correspond to "allowed" obitals. The reason why only these energies are "allowed" forms a significant part of quantum theory (see HL topic 12).

When an electron moves between energy levels it must emit or absorb energy. The energy emitted or absorbed corresponds to the difference between the two allowed energy levels. This energy is emitted or absorbed as "packets" of light called photons (for more information, see option H). A higher energy photon corresponds to a higher frequency (shorter wavelength) of light. Thus the frequency of the light, emitted or absorbed, is fixed by the energy difference between the levels. Since the energy levels are unique to a given element, this means that the emission (and the absorption) spectrum will also be unique.

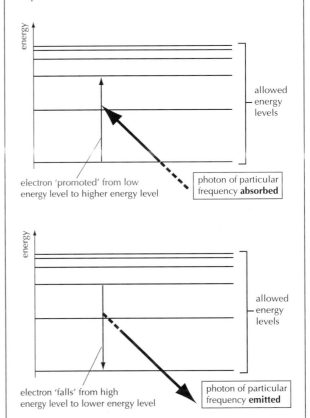

Nuclear structure

ISOTOPES

When a chemical reaction takes place, it involves the outer electrons of the atoms concerned. Different elements have different chemical properties because the arrangement of outer electrons varies from element to element. The chemical properties of a particular element are fixed by the amount of positive charge that exists in the nucleus – in other words, the number of protons. In general, different nuclear structures will imply different chemical properties. A **nuclide** is the name given to a particular species of atom (one whose nucleus contains a specified number of protons and a specified number of neutrons). Some nuclides are the same element – they have the same chemical properties and contain the same number of protons. These nuclides are called **isotopes** – they contain the same number of protons but different numbers of neutrons.

NOTATION

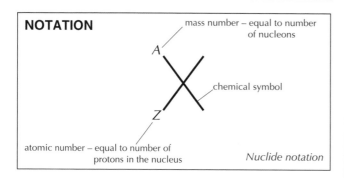

mass number – equal to number of nucleons

chemical symbol

atomic number – equal to number of protons in the nucleus

Nuclide notation

EXAMPLES

	Notation	Description	Comment
1	$^{12}_{6}\text{C}$	carbon-12	isotope of **2**
2	$^{13}_{6}\text{C}$	carbon-13	isotope of **1**
3	$^{238}_{92}\text{U}$	uranium-238	
4	$^{198}_{78}\text{Pt}$	platinum-198	same mass number as **5**
5	$^{198}_{80}\text{Hg}$	mercury-198	same mass number as **4**

Each element has a unique chemical symbol and its own atomic number. *No.1* and *No.2* are examples of two isotopes, whereas *No.4* and *No.5* are not.

In general, when physicists use this notation they are concerned with the nucleus rather than the whole atom. Chemists use the same notation but tend to include the overall charge on the atom. Thus $^{12}_{6}\text{C}$ can represent the carbon nucleus to a physicist or the carbon atom to a chemist depending on the context. If the charge is present the situation becomes unambiguous. $^{35}_{17}\text{Cl}^{-}$ must refer to a chlorine ion – an atom that has gained one extra electron.

STRONG NUCLEAR FORCE

The protons in a nucleus are all positive. Since like charges repel, they must be repelling one another all the time. This means there must be another force keeping the nucleus together. Without it the nucleus would "fly apart". We know a few things about this force.

- It must be strong. If the proton repulsions are calculated it is clear that the gravitational attraction between the nucleons is far too small to be able to keep the nucleus together.
- It must be very short-ranged as we do not observe this force anywhere other than inside the nucleus.
- It is likely to involve the neutrons as well. Small nuclei tend to have equal numbers of protons and neutrons. Large nuclei need proportionately more neutrons in order to keep the nucleus together.

The name given to this force is the **strong nuclear force**.

NUCLEAR STABILITY

Many atomic nuclei are unstable. The stability of a particular nuclide depends greatly on the numbers of neutrons present. The graph below shows the stable nuclides that exist.

- For small nuclei, the number of neutrons tends to equal the number of protons.
- For large nuclei there are more neutrons than protons.
- Nuclides above the band of stability have "too many neutrons" and will tend to decay with either alpha or beta decay. (see page 51)
- Nuclides below the band of stability have "too few neutrons" and will tend to emit positrons. (see page 54)

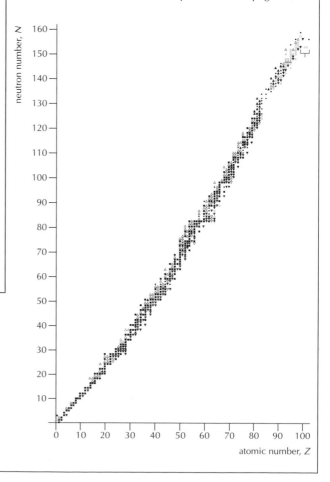

Key

N number of neutrons

Z number of protons

- ■ naturally occurring stable nuclide
- • naturally occurring α-emitting nuclide
- ○ artificially produced α-emitting nuclide
- ▲ naturally occurring β^{-}-emitting nuclide
- △ artificially produced β^{+}-emitting nuclide
- ▽ artificially produced β^{-}-emitting nuclide
- ▼ artificially produced electron-capturing nuclide
- ▼ artificial nuclide decaying by spontaneous fission

Radioactivity

IONISING PROPERTIES

Many atomic nuclei are unstable. The process by which they decay is called **radioactive decay**. Every decay involves the emission of one of three different possible radiations from the nucleus: alpha (α), beta (β) or gamma. (γ). (see also page 54)

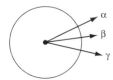

Alpha, beta and gamma all come from the nucleus

All three radiations are ionising. This means that as they go through a substance, collisions occur which cause electrons to be removed from atoms. Atoms that have lost or gained electrons are called ions. This ionising property allows the radiations to be detected. It also explains their dangerous nature. When ionisations occur in biologically important molecules, such as DNA, mutations can occur.

DETECTION

The ionisation chamber and the Geiger-Muller tube ("Geiger counter") can be used to count the number of ionising radiations that enter these detection devices.

They are similar in principle. When radiation enters the chamber it ionises gas molecules. The voltage causes the ions to accelerate towards the electrodes and an electrical pulse passes through the circuit. The number of pulses that arrive per second is a measure of the rate at which ionising radiation is entering the chamber.

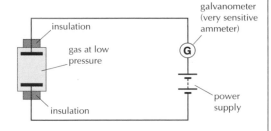

ionization chamber

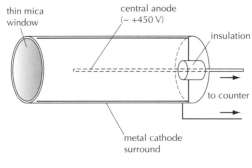

Geiger–Müller tube

Detection devices

PROPERTIES OF ALPHA, BETA, GAMMA

Property	Alpha, α	Beta, β	Gamma, γ
Effect on photographic film	Yes	Yes	Yes
Approximate number of ion-pairs produced in air	10^4 per mm	10^2 per mm	1 per mm
Typical material needed to absorbed it	10^{-2} mm Aluminium piece of paper	A few mm Aluminium	10 cm lead
Penetration ability	Low	Medium	High
Typical path length in air	A few cm	Less than one m	Effectively infinite
Deflection by magnetic and electric fields	Behaves like a positive charge	Behaves like a negative charge	Not deflected
Speed	About 10^7 m s^{-1}	About 10^8 m s^{-1}, very variable	3×10^8 m s^{-1}

NATURE OF ALPHA, BETA AND GAMMA DECAY

When a nucleus decays the mass numbers and the atomic numbers must balance on each side of the nuclear equation.

- Alpha particles are helium nuclei, $^4_2\alpha$ or $^4_2\text{He}^{2+}$. In alpha decay, a "chunk" of the nucleus is emitted. The portion that remains will be a different nuclide.

$$^A_Z X \rightarrow ^{(A-4)}_{(Z-2)} Y + ^4_2\alpha$$

e.g. $^{241}_{95}\text{Am} \rightarrow ^{237}_{93}\text{Np} + ^4_2\alpha$

The atomic numbers and the mass numbers balance on each side of the equation.

$(95 = 93 + 2$ and $241 = 237 + 4)$

- Beta particles are electrons, $^0_{-1}\beta$ or $^0_{-1}e^-$, emitted **from the nucleus**. The explanation is that the electron is formed when a neutron decays. At the same time, another particle is emitted called an antineutrino.

$$^1_0 n \rightarrow ^1_1 P + ^0_{-1}\beta + \bar{\upsilon}$$

Since an antineutrino has no charge and virtually no mass it does not affect the equation and so is sometimes ignored. See page 86 for more details.

$$^A_Z X \rightarrow ^A_{(Z+1)} Y + ^0_{-1}\beta + \bar{\upsilon}$$

e.g. $^{90}_{38}\text{Sr} \rightarrow ^{90}_{90}Y + ^0_{-1}\beta + \bar{\upsilon}$

- Gamma rays are unlike the other two radiations in that they are part of the electromagnetic spectrum. After their emission, the nucleus has less energy but its mass number and its atomic number have not changed. It is said to have changed from an **excited state** to a lower energy state.

$$^A_Z X^* \rightarrow ^A_Z X + ^0_0\gamma$$

Excited state   Lower energy state

Half-life

RANDOM DECAY

Radioactive decay is a **random** process and is not affected by external conditions. For example, increasing the temperature of a sample of radioactive material does not affect the rate of decay. This means that is there no way of knowing whether or not a particular nucleus is going to decay within a certain period of time. All we know is the *chances* of a decay happening in that time.

Although the process is random, the large numbers of atoms involved allows us to make some accurate predictions. If we start with a given number of atoms then we can expect a certain number to decay in the next minute. If there were more atoms in the sample, we would expect the number decaying to be larger. On average the rate of decay of a sample is proportional to the number of atoms in the sample. This proportionality means that radioactive decay is an **exponential** process. The number of atoms of a certain element, N, decreases exponentially over time. Mathematically this is expressed as:

$$\frac{dN}{dt} \propto -N$$

HALF-LIFE

There is a temptation to think that every quantity that decreases with time is an exponential decrease, but exponential curves have a particular mathematical property. In the graph shown below, the time taken for half the number of nuclides to decay is always the same, whatever starting value we choose. This allows us to express the chances of decay happening in a property called the **half-life,** $T_{1/2}$. The half-life of a nuclide is the time taken for half the number of nuclides present in a sample to decay. An equivalent statement is that the half-life is the time taken for the rate of decay of a particular sample of nuclides to halve. A substance with a large half-life takes a long time to decay. A substance with a short half-life will decay quickly. Half-lives can vary from fractions of a second to millions of years.

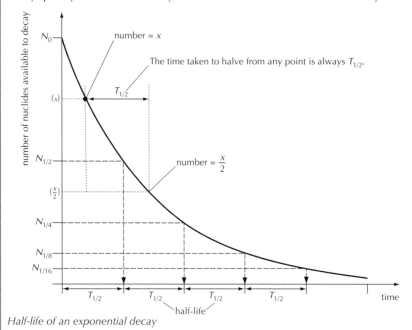

Half-life of an exponential decay

EXAMPLE

In simple situations, working out how much radioactive material remains is a matter of applying the half-life property several times. A common mistake is to think that if the half-life of a radioactive material is 3 days then it will all decay in six days. In reality, after six days (two half-lives) a "half of a half" will remain i.e. a quarter.

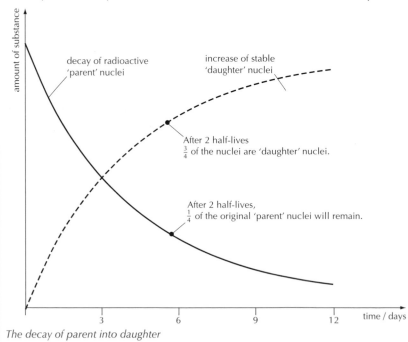

The decay of parent into daughter

e.g. The half-life of $^{14}_{6}C$ is 5570 years. Approximately how long is needed before less than 1% of a sample of $^{14}_{6}C$ remains?

Time	Fraction left
$T_{1/2}$	50%
$2T_{1/2}$	25%
$3T_{1/2}$	12.5%
$4T_{1/2}$	~ 6.3%
$5T_{1/2}$	~ 3.1%
$6T_{1/2}$	~ 1.5%
$7T_{1/2}$	~ 0.8%

6 half lives	= 33420 years
7 half lives	= 38990 years

∴ approximately 37000 years needed

Nuclear reactions

ARTIFICIAL TRANSMUTATIONS

There is nothing that we can do to change the likelihood of a certain radioactive decay happening, but under certain conditions we can make nuclear reactions happen. This can be done by bombarding a nucleus with a nucleon, an alpha particle or another small nucleus. Such reactions are called **artificial transmutations**. In general, the target nucleus first "captures" the incoming object and then an emission takes place. The first ever artificial transmutation was carried out by Rutherford in 1919. Nitrogen was bombarded by alpha particles and the presence of oxygen was detected spectroscopically.

$$_2^4He^{2+} + {}_7^{14}N \rightarrow {}_8^{17}O + {}_1^1p$$

The mass numbers (4 + 14 = 17 +1) and the atomic numbers (2 + 7 = 8 +1) on both sides of the equation must balance.

UNIFIED MASS UNITS

The individual masses involved in nuclear reactions are tiny. In order to compare atomic masses physicists often use unified mass units, u. These are defined in terms of the most common isotope of carbon, carbon-12. There are 12 nucleons in the carbon-12 atom (6 protons and 6 neutons) and one unified mass unit is defined as exactly one twelfth the mass of a carbon-12 atom. Essentially, the mass of a proton and the mass of a neutron are both 1 u as shown in the table below.

$$1 \text{ u} = \frac{1}{12} \text{ mass of a (carbon-12) atom} = 1.66 \times 10^{-27} \text{ kg}$$

mass* of 1 proton = 1.007 276 u

mass* of 1 neutron = 1.008665 u

mass* of 1 electron = 0.000549 u

* = Technically these are all "rest masses" – see Relativity option

MASS DEFECT AND BINDING ENERGY

The table above shows the masses of neutrons and protons. It should be obvious that if we add together the masses of 6 protons, 6 neutrons and 6 electrons we will get a number bigger than 12 u, the mass of a carbon-12 atom. What has gone wrong? The answer becomes clear when we investigate what keeps the nucleus bound together.

The difference between the mass of a nucleus and the masses of its component nucleons is called the **mass defect**. If one imagined assembling a nucleus, the protons and neutrons would initially need to be brought together. Doing this takes work because the protons repel one another. The energy needed to do this work must come from somewhere. The answer lies in Einstein's famous mass-energy equivalence relationship.

$$E = mc^2$$

energy in joules   mass in kg   speed of light in m s^{-1}

In Einstein's equation, mass is another form of energy and it is possible to convert mass directly into energy and vice versa. The **binding energy** is the amount of energy that is released when a nucleus is assembled from its component nucleons. It comes from a decrease in mass. The binding energy would also be the energy that needs to be added in order to separate a nucleus into its individual nucleons. The mass defect is thus a measure of the binding energy.

UNITS

Using Einstein's equation, 1 kg of mass is equivalent to 9×10^{16} J of energy. This is a huge amount of energy. At the atomic scale other units of energy tend to be more useful. The electronvolt (see topic 5), or more usually, the megaelectronvolt are often used.

$$1 \text{ eV} = 1.6 \times 10^{-19} \text{ J}$$

$$1 \text{ MeV} = 1.6 \times 10^{-13} \text{ J}$$

$$1 \text{ u of mass converts into 935.5 MeV}$$

Since mass and energy are equivalent it is sometimes useful to work in units that avoid having to do repeated multiplications by the (speed of light)^2. A new possible unit for mass is thus MeV c^{-2}. It works like this:

If 1 MeV c^{-2} worth of mass is converted you get 1 MeV worth of energy.

WORKED EXAMPLES

Question:

How much energy would be released if 14 g of carbon-14 decayed as shown in the equation below?

$$_6^{14}C \rightarrow {}_7^{14}N + {}_{-1}^0\beta$$

Answer:

Information given

atomic mass of carbon-14 = 14.003242 u;

atomic mass of nitrogen-14 = 14.003074 u;

mass of electron = 0.000549 u

mass of left-hand side = nuclear mass of $_6^{14}C$

$$= 14.003242 - 6(0.000549) \text{ u}$$

$$= 13.999948 \text{ u}$$

nuclear mass of $_7^{14}N = 14.003074 - 7(0.000549)$ u

$$= 13.999231 \text{ u}$$

mass of right-hand side = 13.999231 + 0.000549 u

$$= 13.999780 \text{ u}$$

mass difference = LHS – RHS

$$= 0.000168 \text{ u}$$

energy released per decay = 0.000168 × 935.5 MeV

$$= 0.157164 \text{ MeV}$$

14g of C-14 is 1 mol

∴ Total number of decays = $N_A = 6.22 \times 10^{23}$

∴ Total energy release = $6.022 \times 10^{23} \times 0.157164$ MeV

$$= 9.464 \times 10^{22} \text{ MeV}$$

$$= 15143 \text{ J}$$

$$\approx 15 \text{ kJ}$$

Fission, fusion and antimatter

FISSION

Fission is the name given to the nuclear reaction whereby large nuclei are induced to break up into smaller nuclei and release energy in the process. It is the reaction that is used in nuclear reactors and atomic bombs. A typical single reaction might involve bombarding a uranium nucleus with a neutron. This can cause the uranium nucleus to break up into two smaller nuclei. A typical reaction might be:

$$^{1}_{0}n + ^{235}_{92}U \rightarrow ^{141}_{56}Ba + ^{92}_{36}Kr + 3 ^{1}_{0}n + energy$$

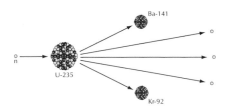

A fission reaction

Since the one original neutron causing the reaction has resulted in the production of three neutrons, there is the

possibility of a **chain reaction** occurring. It is technically quite difficult to get the neutrons to lose enough energy to go on and initiate further reactions, but it is achievable.

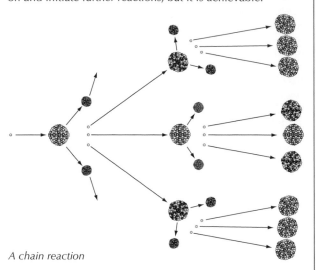

A chain reaction

FUSION

Fusion is the name given to the nuclear reaction whereby small nuclei are induced to join together into larger nuclei and release energy in the process. It is the reaction that "fuels" all stars including the Sun. A typical reaction that is taking place in the Sun is the fusion of two different isotopes of hydrogen to produce helium.

$$^{2}_{1}H + ^{3}_{1}H \rightarrow ^{4}_{2}He + ^{1}_{0}n + energy$$

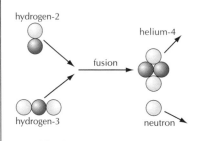

One of the fusion reactions happening in the Sun

BINDING ENERGY PER NUCLEON

Whenever a nuclear reaction (fission or fusion) releases energy, the products of the reaction are in a lower energy state than the reactants. Mass loss must be the source of this energy. In order to compare the energy states of different nuclei, physicists calculate the binding energy per nucleon. This is the total binding energy for the nucleus divided by the total number of nucleons. The nucleus with the largest binding energy per nucleon is iron-56, $^{56}_{26}Fe$.

A reaction is energetically feasible if the products of the reaction have a greater binding energy per nucleon when compared with the reactants.

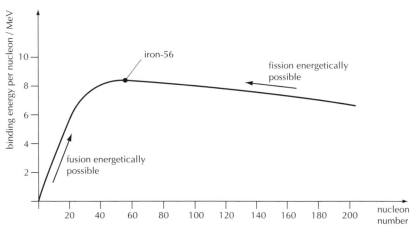

Graph of binding energy per nucleon number

ANTIMATTER

The nuclear model given in the previous pages is somewhat simplified but is all that is required for the IB Standard level examination. One important thing that has not been mentioned is the existence of antimatter. Every form of matter has its equivalent form of antimatter. If matter and antimatter came together they would annihilate each other. Not surprisingly, antimatter is rare but it does exist. For example, another form of radioactive decay that can take place is beta plus or positron decay. In this decay a proton

decays into a neutron, and the antimatter version of an electron, a positron, is emitted.

$$^{1}_{1}p \rightarrow ^{1}_{0}n + ^{0}_{+1}\beta^{+} + \upsilon$$

The positron, β^{+}, emission is accompanied by a neutrino.

The antineutrino is the antimatter form of the neutrino.

$$^{19}_{10}Ne \rightarrow ^{19}_{9}F + ^{0}_{+1}\beta^{+} + \bar{\upsilon}$$

For more details see the Higher level material in section 12.

1 A sample of radioactive material contains the element Ra 226. The half-life of Ra 226 can be defined as the time it takes for

A the mass of the sample to fall to half its original value.

B half the number of atoms of Ra 226 in the sample to decay.

C half the number of atoms in the sample to decay.

D the volume of the sample to fall to half its original value.

2 Oxygen-15 decays to nitrogen-15 with a half-life of approximately 2 minutes. A pure sample of oxygen-15, with a mass of 100 g, is placed in an airtight container. After 4 minutes, the masses of oxygen and nitrogen in the container will be

	Mass of oxygen	Mass of nitrogen
A	0 g	100 g
B	25 g	25 g
C	50 g	50 g
D	25 g	75 g

3 A radioactive nuclide $_ZX$ undergoes a sequence of radioactive decays to form a new nuclide $_{Z+2}Y$. The sequence of emitted radiations could be

A β, β **B** α, β, β **C** α, α **D** α, β, γ

4 In the Rutherford scattering experiment, a stream of α particles is fired at a thin gold foil. Most of the α particles

A are scattered randomly.

B rebound.

C are scattered uniformly.

D go through the foil.

5 A piece of radioactive material now has about 1/16 of its previous activity. If the half-life is 4 hours the difference in time between measurements is approximately

A 8 hours. **B** 16 hours. **C** 32 hours. **D** 60 hours.

6 The nuclide $_6^{14}C$ undergoes radioactive beta-minus decay. The resulting daughter nuclide is

A $_6^{14}C$ **B** $_6^{10}Be$ **C** $_7^{14}N$ **D** $_5^{14}B$

7 Although there are protons in close proximity to each other in a nucleus of an atom, the nucleus does not blow apart by electrostatic Coulomb repulsion because

A there are an equal number of electrons in the nucleus which neutralise the protons.

B the Coulomb force does not operate in a nucleus.

C the neutrons in the nucleus shield the protons from each other.

D there is a strong nuclear force which counteracts the repulsive Coulomb force.

8 (a) Two properties of the isotope of Uranium, $_{92}^{238}U$ are:
  (i) it decays radioactively (to $_{90}^{234}Th$)
  (ii) it reacts chemically (e.g. with Fluorine to form UF_6).

What features of the structure of Uranium atoms are responsible for these two widely different properties? [2]

(b) A beam of deuterons (Deuterium nuclei, $_1^2H$) are accelerated through a potential difference and are then incident on a Magnesium target ($_{12}^{26}Mg$). A nuclear reaction occurs resulting in the production of a Sodium nucleus and an alpha particle.
  (i) Write a balanced nuclear equation for this reaction. [2]
  (ii) Explain why it is necessary to give the deuterons a certain minimum kinetic energy before they can react with the magnesium nuclei. [2]

9 *Radioactive carbon dating*

The carbon in trees is mostly carbon-12, which is stable, but there is also a small proportion of carbon-14, which is radioactive. When a tree is cut down, the carbon-14 present in the wood at that time decays with a half-life of 5800 years.

(a) Carbon-14 decays by beta-minus emission to nitrogen-14. Write the equation for this decay. [2]

(b) For an old wooden bowl from an archaeological site, the average count-rate of beta particles detected per kg of carbon is 13 counts per minute. The corresponding count rate from newly cut wood is 52 counts per minute.
  (i) Explain why the beta activity from the bowl diminishes with time, even though the probability of decay of any individual carbon-14 nucleus is constant. [3]
  (ii) Calculate the approximate age of the wooden bowl. [3]

10 This question is about a nuclear fission reactor for providing electrical power.

In a nuclear reactor, power is to be generated by the fission of uranium-235. The absorption of a neutron by ^{235}U results in the splitting of the nucleus into two smaller nuclei plus a number of neutrons and the release of energy. The splitting can occur in many ways; for example

$n + _{92}^{235}U \rightarrow _{38}^{90}Sr + _{54}^{143}Xe + \text{neutrons} + \text{energy}$

(a) *The nuclear fission reaction*
  (i) How many neutrons are produced in this reaction? [1]
  (ii) Explain why the release of several neutrons in each reaction is crucial for the operation of a fission reactor. [2]
  (iii) The sum of the rest masses of the uranium plus neutron before the reaction is 0.22 u greater than the sum of the rest masses of the fission products. What becomes of this 'missing mass'? [1]
  (iv) Show that the energy released in the above fission reaction is about 200 MeV. [2]

(b) *A nuclear fission power station*
  (i) Suppose a nuclear fission power station generates electrical power at 550 MW. Estimate the minimum number of fission reactions occurring each second in the reactor, stating any assumption you have made about efficiency. [4]

Graphical analysis – Logarithmic functions

LOGS – BASE TEN AND BASE e

Mathematically,

If $a = 10^b$

Then $\log(a) = b$ [to be absolutely precise $\log_{10}(a) = b$]

Most calculators have a 'log' button on them. But we don't **have** to use 10 as the base. We can use any number that we like. For example we could use 2.0, 563.2, 17.5, 42 or even 2.7182818284590452353602874714. For complex reasons this last number IS the most useful number to use! It is given the symbol e and logarithms to this base are called **natural logarithms**. The symbol for natural logarithms is $\ln(x)$. This is also on most calculators.

If $p = e^q$

Then $\ln(p) = q$

The powerful nature of logarithms means that we have the following rules

$\ln(c \times d) = \ln(c) + \ln(d)$

$\ln(c \div d) = \ln(c) - \ln(d)$

$\ln(c^n) = n\ln(c)$

$\ln\left(\frac{1}{c}\right) = -\ln(c)$

These rules have been expressed for natural logarithms, but they work for all logarithms whatever the base.

The point of logarithms is that they can be used to express some relationships (particularly power laws and exponentials) in straight-line form. This means that we will be plotting graphs with logarithmic scales.

A normal scale increases by the same amount each time.

1 2 3 4 5 6 7 8 9 10 11

A logarithmic scale increases by the same ratio all the time.

10^0 10^1 10^2 10^3
1 10 100 1000

EXPONENTIALS AND LOGS (LOG-LINEAR)

Natural logarithms are very important because many natural processes are exponential. Radioactive decay is an important example. In this case, once again the taking of logarithms allows the equation to be compared with the equation for a straight-line.

For example, the count rate R at any given time t is given by the equation

$R = R_0\, e^{-\lambda t}$

R_0 and λ are constants.

POWER LAWS AND LOGS (LOG-LOG)

When an experimental situation involves a power law – it is often only possible to transform it into straight-line form by taking logs. For example, the time period of a simple pendulum, T, is related to its length, l, by the following equation.

$T = k\, l^p$

k and p are constants.

A plot of the variables will give a curve, but it is not clear from this curve what the values of k and p work out to be. On top of this, if we do not know what the value of p is, we can not calculate the values to plot a straight-line graph.

Time period versus length for a simple pendulum

The 'trick' is to take logs of both sides of the equation.

$\ln(T) = \ln(k\, l^p)$

$\ln(T) = \ln(k) + \ln(l^p)$

$\ln(T) = \ln(k) + p\ln(l)$

This is now in the same form as the equation for a straight-line

$y = c + mx$

If we take logs, we get

$\ln(R) = \ln(R_0\, e^{-\lambda t})$

$\ln(R) = \ln(R_0) + \ln(e^{-\lambda t})$

$\ln(R) = \ln(R_0) - \lambda t\ln(e)$

$\ln(R) = \ln(R_0) - \lambda t$ [$\ln(e) = 1$]

This can be compared with the equation for a straight-line graph

$y = c + mx$

Thus if we plot $\ln(R)$ on the y-axis and t on the x-axis, we will get a straight line.

Gradient $= -\lambda$

Intercept $= \ln(R_0)$

Thus if we plot $\ln(T)$ on the y-axis and $\ln(l)$ on the x-axis we will get a straight-line graph.

The gradient will be equal to p

The intercept will be equal to $\ln(k)$ [so $k = e^{(\text{intercept})}$]

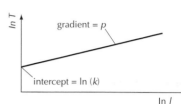

Plot of ln (time period) versus ln (length) gives a straight-line graph.

Both the gravity force and the electrostatic force are inverse-square relationships. This means that the force $\propto$ (distance apart)$^{-2}$. The same technique can be used to generate a straight-line graph.

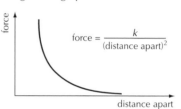

$\text{force} = \dfrac{k}{(\text{distance apart})^2}$

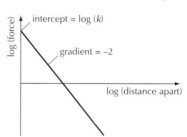

Inverse square relationship – direct plot and log-log plot

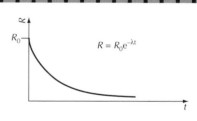

$R = R_0 e^{-\lambda t}$

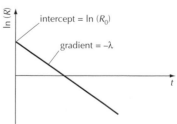

Exponential decrease – direct plot and log-linear plot

HL Uncertainties in calculated results

MATHEMATICAL REPRESENTATION OF UNCERTAINTIES

For example if the mass of a block was measured as 10 ± 1 g and the volume was measured as 5.0 ± 0.2 cm^3, then the full calculations for the density would be as follows.

Best value for density $= \dfrac{\text{mass}}{\text{volume}} = \dfrac{10}{5} = 2.0$ g cm^{-3}

The largest possible value of density $= \dfrac{11}{4.8} = 2.292$ g cm^{-3}

The smallest possible value of density $= \dfrac{9}{5.2} = 1.731$ g cm^{-3}

Rounding these values gives density $= 2.0 \pm 0.3$ g cm^{-3}

We can express this uncertainty in one of three ways – using **absolute**, **fractional** or **percentage uncertainties**.

If a quantity p is measured then the absolute uncertainty would be expressed as $\pm \Delta p$.

Then the fractional uncertainty is
$$\frac{\pm \Delta p}{p},$$
which makes the percentage uncertainty
$$\frac{\pm \Delta p}{p} \times 100\%.$$

In the example above, the uncertainty is ± 0.15 or $\pm 15\%$.

Thus equivalent ways of expressing this error are
density $= 2.0 \pm 0.3$ g cm^{-3}
OR density $= 2.0$ g cm$^{-3} \pm 15\%$

Working out the uncertainty range is very time consuming. There are some mathematical 'short-cuts' that can be used. These are introduced in the boxes below.

MULTIPLICATION, DIVISION OR POWERS

Whenever two or more quantities are multiplied or divided and they each have uncertainties, the overall uncertainty is approximately equal to the **addition** of the **percentage** (fractional) uncertainties.

Using the same numbers from above,

$\Delta m = \pm 1$ g

$\dfrac{\Delta m}{m} = \pm \left(\dfrac{1 \text{ g}}{10 \text{ g}}\right) = \pm 0.1 = \pm 10\%$

$\Delta v = \pm 0.2$ cm^3

$\dfrac{\Delta V}{V} = \pm \left(\dfrac{0.2 \text{ cm}^3}{5 \text{ cm}^3}\right) = \pm 0.04 = \pm 4\%$

The total % uncertainty in the result $= \pm(10 + 4)\%$
$= \pm 14\%$

14% of 2.0 g cm^{-3} = 0.28 g cm^{-3} ≈ 0.3 g cm^{-3}
So density $= 2.0 \pm 0.3$ g cm^{-3} as before.

In symbols, if $P = Q \times R$ or if $P = \dfrac{Q}{R}$

Then $\dfrac{\Delta P}{P} = \dfrac{\Delta Q}{Q} + \dfrac{\Delta R}{R}$ [note this is ALWAYS added]

Power relationships are just a special case of this law.
If $P = R^n$

Then $\dfrac{\Delta P}{P} = n\left(\dfrac{\Delta R}{R}\right)$

For example if a cube is measured to be 4.0 ± 0.1 cm in length along each side, then

% Uncertainty in length $= \pm \left(\dfrac{0.1}{4.0}\right) = \pm 2.5\%$

Volume = (length)^3 = $(4.0)^3$ = 64 cm^3

% Uncertainty in [volume] = % uncertainty in [(length)^3]
$= 3 \times$ (% uncertainty in [length])
$= 3 \times (\pm 2.5\%)$
$= \pm 7.5\%$

Absolute uncertainty = 7.5% of 64 cm^3
$= 4.8$ cm^3 ≈ 5 cm^3

Thus volume of cube = 64 ± 5 cm^3

OTHER MATHEMATICAL OPERATIONS

If the calculation involves mathematical operations other than multiplication, division or raising to a power, then one has to find the highest and lowest possible values.

Addition or subtraction

Whenever two or more quantities are added or subtracted and they each have uncertainties, the overall uncertainty is equal to the **addition** of the **absolute** uncertainties.

uncertainty of thickness in a pipe wall

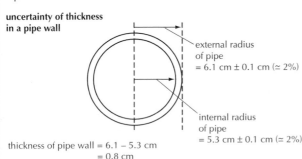

external radius of pipe = 6.1 cm ± 0.1 cm (≈ 2%)
internal radius of pipe = 5.3 cm ± 0.1 cm (≈ 2%)

thickness of pipe wall = 6.1 – 5.3 cm
$= 0.8$ cm

uncertainty in thickness $= \pm(0.1 + 0.1)$ cm
$= 0.2$ cm
$= \pm 25\%$

Other functions

There are no 'short-cuts' possible.

uncertainty of sin θ if $\theta = 60° \pm 5°$

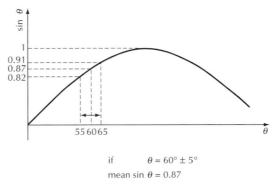

if $\theta = 60° \pm 5°$
mean sin $\theta = 0.87$
max. sin $\theta = 0.91$
min. sin $\theta = 0.82$
∴ sin $\theta = 0.87 \pm 0.05$

worst value used

HL Uncertainties in graphs

ERROR BARS

Plotting a graph allows one to visualise all the readings at one time. Ideally all of the points should be plotted with their error bars. In principle, the size of the error bar could well be different for every single point and so they should be individually worked out.

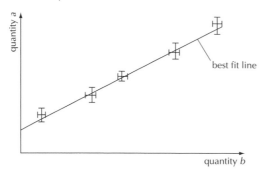

In practice, it would often take too much time to add all the correct error bars, so some (or all) of the following short cuts could be considered.

- Rather than working out error bars for each point – use the worst value and assume that all of the other error bars are the same.
- Only plot the error bar for the 'worst' point i.e. the point that is furthest from the line of best fit. If the line of best fit is within the limits of this error bar, then it will probably be within the limits of all the error bars.
- Only plot the error bars for the first and the last points. These are often the most important points when considering the uncertainty ranges calculated for the gradient or the intercept (see below).
- Only include the error bars for the axis that has the worst uncertainty.

UNCERTAINTY IN SLOPES

If the gradient of the graph has been used to calculate a quantity, then the uncertainties of the points will give rise to an uncertainty in the gradient. Using the steepest and the shallowest lines possible (i.e. the lines that are still consistent with the error bars) the uncertainty range for the gradient is obtained. This process is represented below.

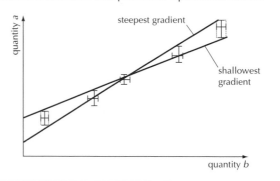

UNCERTAINTY IN INTERCEPTS

If the intercept of the graph has been used to calculate a quantity, then the uncertainties of the points will give rise to an uncertainty in the intercept. Using the steepest and the shallowest lines possible (i.e. the lines that are still consistent with the error bars) we can obtain the uncertainty in the result. This process is represented below.

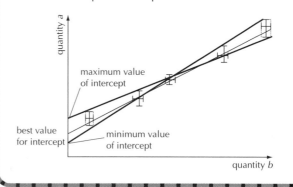

1 The lengths of the sides of a rectangular plate are measured, and the diagram shows the measured values with their uncertainties.

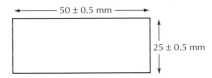

- 50 ± 0.5 mm -

25 ± 0.5 mm

Which one of the following would be the best estimate of the percentage uncertainty in the calculated area of the plate?

A ± 0.02%   C ± 3%

B ± 1%   D ± 5%

2 A stone is dropped down a well and hits the water 2.0 s after it is released. Using the equation $d = \frac{1}{2}g\,t^2$ and taking $g = 9.81$ m s^{-2}, a calculator yields a value for the depth d of the well as 19.62 m. If the time is measured to ±0.1 s then the best estimate of the absolute error in d is

A ±0.1 m   C ±1.0 m

B ±0.2 m   D ±2.0 m

3 In order to determine the density of a certain type of wood, the following measurements were made on a **cube** of the wood.

    Mass   = 493 g
    Length of **each** side = 9.3 cm

The percentage uncertainty in the measurement of mass is ±0.5% and the percentage uncertainty in the measurement of length is ±1.0%.

The best estimate for the uncertainty in the density is

A ±0.5%   C ±3.0%

B ±1.5%   D ±3.5%

4 The graphs A to D below are plots of log y against log x in arbitrary units.

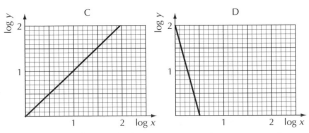

Which one of the graphs best represents the variation of y, the electrostatic potential due to a positive point charge, with x, the distance from the point charge?

5 This question is about finding the relationship between the forces between magnets and their separations.

In an experiment, two magnets were placed with their North-seeking poles facing one another. The force of repulsion, f, and the separation of the magnets, d, were measured and the results are shown in the table below.

Separation d/m	Force of repulsion f/N
0.04	4.00
0.05	1.98
0.07	0.74
0.09	0.32

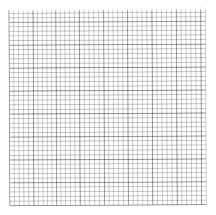

(a) Plot a graph of log (force) against log (distance). [3]

(b) The law relating the force to the separation is of the form

$$f = kd^n$$

(i) Use the graph to find the value of n. [2]

(ii) Calculate a value for k, giving its units. [3]

6 Astronauts wish to determine the gravitational acceleration on Planet X by dropping stones from an overhanging cliff. Using a steel tape measure they measure the height of the cliff as $s = 7.64$ m ± 0.01 m. They then drop three similar stones from the cliff, timing each fall using a hand-held electronic stopwatch which displays readings to one-hundredth of a second. The recorded times for three drops are 2.46 s, 2.31 s and 2.40 s.

(a) Explain why the time readings vary by more than a tenth of a second, although the stopwatch gives readings to one hundredth of a second. [1]

(b) Obtain the average time t to fall, and write it in the form (value ± uncertainty), to the appropriate number of significant digits. [1]

(c) The astronauts then determine the gravitational acceleration a_g on the planet using the formula $a_g = \frac{2s}{t^2}$. Calculate a_g from the values of s and t, and determine the uncertainty in the calculated value. Express the result in the form a_g = (value ± uncertainty), to the appropriate number of significant digits. [3]

Projectile motion

COMPONENTS OF PROJECTILE MOTION

If two children are throwing and catching a tennis ball between them, the path of the ball is always the same shape. This motion is known as **projectile motion** and the shape is called a **parabola**.

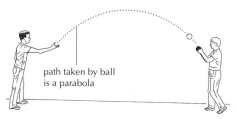

path taken by ball
is a parabola

The only forces acting during its flight are gravity and friction. In many situations, air resistance can be ignored.

It is moving horizontally and vertically **at the same time** but the horizontal and vertical components of the motion are **independent** of one another. Assuming the gravitional force is constant, this is always true.

Horizontal component

There are no forces in the horizontal direction, so there is no horizontal acceleration. This means that the horizontal velocity must be constant.

ball travels at a constant horizontal velocity

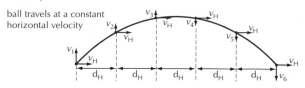

Vertical component

There is a constant vertical force acting down, so there is a constant vertical acceleration. The value of the vertical acceleration is 10 m s^{-2} – the acceleration due to gravity.

vertical velocity changes

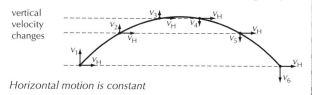

Horizontal motion is constant

MATHEMATICS OF PARABOLIC MOTION

The graphs of the components of parabolic motion are shown below.

motion in the x-direction

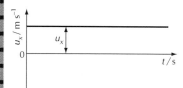

motion in the y-direction

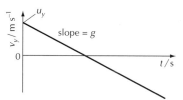

slope = g

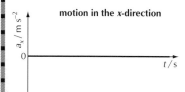

slope = u_x

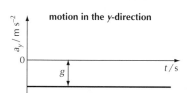

maximum height

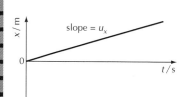

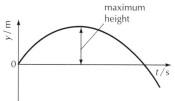

Once the components have been worked out, the actual velocities (or displacements) at any time can be worked out by vector addition.

The solution of any problem involving projectile motion is as follows:
• use the angle of launch to resolve the initial velocity into components.
• the time of flight will be determined by the vertical component of velocity.
• the range will be determined by the horizontal component (and the time of flight).
• the velocity at any point can be found by vector addition.

Useful 'shortcuts' in calculations include the following facts:
• for a given speed, the greatest range is achieved if the launch angle is 45°.
• if two objects are released together, one with a horizontal velocity and one from rest, they will both hit the ground together.

EXAMPLE

A projectile is launched horizontally from the top of a cliff.

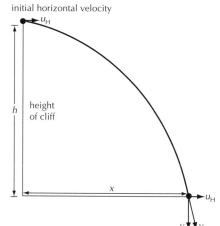

initial horizontal velocity

height of cliff

vertical motion	horizontal motion
$u = 0$	$u = u_H$
$v = ?$	$v = u_H$
$a = 10$ m s^{-2}	$a = 0$
$s = h$	$s = x$
$t = ?$	$t = ?$

$$s = ut + \frac{1}{2}at^2$$

so $\quad h = 0 + \frac{1}{2} \times 10 \times t^2$

∴ $\quad t^2 = \frac{2h}{10}$

$\quad t = \sqrt{\frac{2h}{10}}$ s

Since $\quad v = u + at$

$v = 0 + 10\sqrt{\frac{2h}{10}}$ m $\qquad x = u_H \times t$

$\quad = \sqrt{20h}$ m s^{-1} $\qquad = u_H \times \sqrt{\frac{2h}{10}}$ m

The final velocity v_f is the vector addition of v and u_H.

Gravitational force and field

NEWTON'S LAW OF UNIVERSAL GRAVITATION

If you trip over, you will fall down towards the ground.

Newton's theory of **universal gravitation** explains what is going on. It is called 'universal' gravitation because at the core of this theory is the statement that every mass in the Universe attracts all the other masses in the Universe. The value of the attraction between two **point** masses is given by an equation.

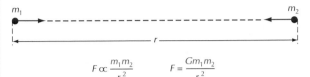

$$F \propto \frac{m_1 m_2}{r^2} \qquad F = \frac{G m_1 m_2}{r^2}$$

Universal gravitational constant $G = 6.67 \times 10^{-11}$ Nm² kg⁻²

The following points should be noticed:
- the law only deals with point masses.
- technically speaking, the masses in this equation are gravitational masses (as opposed to inertial masses – see page 17 for more details).
- there is a force acting on each of the masses. These forces are EQUAL and OPPOSITE (even if the masses are not equal).
- the forces are always attractive.
- gravitation forces act between ALL objects in the Universe. The forces only become significant if one (or both) of the objects involved are massive, but they are there nonetheless.

The interaction between two spherical masses turns out to be the same as if the masses were concentrated at the centres of the spheres.

GRAVITATIONAL FIELD STRENGTH

The table below should be compared with the one on page 44.

	Gravitational field strength
Symbol	g
Caused by...	Masses
Affects...	Masses
One type of...	Mass
Simple force rule:	All masses attract

The gravitational field is therefore defined as the force per unit mass. $\mathbf{g} = \dfrac{F}{m}$ m = test mass

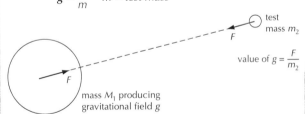

value of $g = \dfrac{F}{m_2}$

mass M_1 producing gravitational field g

The SI units for g are N kg⁻¹. These are the same as m s⁻². Field strength is a vector quantity and can be represented by the use of field lines.

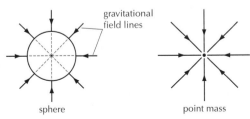

Field strength around masses (sphere and point)

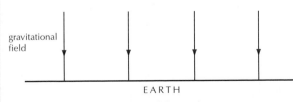

Gravitational field near surface of the Earth

In the examples left the numerical value for the gravitational field can be calculated using Newton's law:

$$g = \frac{G M}{r^2}$$

The gravitational field strength at the surface of a planet must be the same as the acceleration due to gravity on the surface.

Field strength is defined to be $\dfrac{\text{force}}{\text{mass}}$.

Acceleration = $\dfrac{\text{force}}{\text{mass}}$ (from $F = ma$)

For the Earth

$M = 6.0 \times 10^{24}$ kg

$r = 6.4 \times 10^3$ m

$$g = \frac{6.67 \times 10^{-11} \times 6.0 \times 10^{24}}{(6.4 \times 10^3)^2} = 9.8 \text{ ms}^{-2}$$

EXAMPLE

In order to calculate the overall gravitational field strength at any point we must use vector addition. The overall gravitational field strength at any point between the Earth and the Moon must be a result of both pulls.

There will be a single point somewhere between the Earth and the Moon where the total gravitational field due to these two masses is zero. Up to this point the overall pull is back to the Earth, after this point the overall pull is towards the Moon.

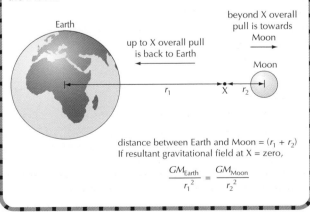

distance between Earth and Moon = $(r_1 + r_2)$
If resultant gravitational field at X = zero,

$$\frac{GM_{\text{Earth}}}{r_1^2} = \frac{GM_{\text{Moon}}}{r_2^2}$$

HL Gravitational potential energy and potential

GRAVITATIONAL POTENTIAL ENERGY

It is easy to work out the difference in gravitational energy when a mass moves between two different heights near the Earth's surface.

The difference in energies $= m\,g\,h$

There are two important points to note:

- this derivation has assumed that the gravitational field strength g is constant. However, Newton's theory of universal gravitation states that the field MUST CHANGE with distance. **This equation can only be used if the vertical distance we move is not very large.**
- the equation assumes that the gravitational potential energy gives zero PE at the surface of the Earth. This works for everyday situations but it is not fundamental.

The true zero of gravitational potential energy is taken at infinity.

If the potential energy of the mass, m, was zero at infinity, and it lost potential energy moving in towards mass M, the potential energy must be **negative** at a given point, P.

The value of gravitational potential energy of a mass at any point in space is defined as the work done in moving it from infinity to that point. The mathematics needed to work this out is not trivial since the force changes with distance.

It turns out that

Gravitational potential energy of mass $m = -\dfrac{G\,M\,m}{r}$
(due to M)

This is a scalar quantity (measured in joules) and is independent of the path taken from infinity.

potential energy decreases as gravitational force does work

zero of potential energy taken to be at infinity

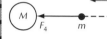

as m moves towards M in the force on m increases

GRAVITATIONAL POTENTIAL

We can define a new quantity, the **gravitational potential** V_g that measures the energy per unit test mass.

$$V_g = \frac{W}{m} \quad \frac{(work\ done)}{(test\ mass)}$$

The SI units of gravitational potential are J kg^{-1}. It is a scalar quantity.

Using Newton's law of universal gravitation, we can work out the gravitational potential at a distance r from any point mass.

$$V_g = -\frac{GM}{r}$$

This formula and the graph also work for spherical masses (planets etc.). The gravitational potential as a result of lots of masses is just the addition of the individual potentials. This is an easy sum since potential is a scalar quantity.

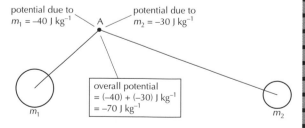

potential due to $m_1 = -40$ J kg^{-1}
potential due to $m_2 = -30$ J kg^{-1}

overall potential
$= (-40) + (-30)$ J kg^{-1}
$= -70$ J kg^{-1}

Once you have the potential at one point and the potential at another, the difference between them is the energy you need to move a unit mass between the two points. It is independent of the path taken.

ESCAPE SPEED

The escape speed of a rocket is the speed needed to be able to escape the gravitational attraction of the planet. This means getting to an infinite distance away.

We know that gravitational potential at the surface of a planet $= \dfrac{-GM}{R_p}$.

(where R_p is the radius of the planet)

This means that for a rocket of mass m, the difference between its energy at the surface and at infinity $= \dfrac{GMm}{R_p}$

Therefore the minimum kinetic energy needed $= \dfrac{GMm}{R_p}$

In other words,

$$\frac{1}{2}m\,(v_{escape})^2 = \frac{GMm}{R_p}$$

so

$$v_{escape} = \sqrt{\left(\frac{2\,GM}{R_p}\right)}$$

This derivation assumes the planet is isolated.

EXAMPLE

The escape speed from an isolated planet like Earth (radius of Earth $R_E = 6.37 \times 10^6$ m) is calculated as follows:

$$v_{escape} = \sqrt{\left(\frac{2 \times 6.67 \times 10^{-11} \times 5.98 \times 10^{24}}{6.37 \times 10^6}\right)} \text{ m s}^{-1}$$

$$= \sqrt{(1.25 \times 10^8)} \text{ m s}^{-1}$$

$$= 1.12 \times 10^4 \text{ m s}^{-1}$$

$$\approx 11 \text{ km s}^{-1}$$

The vast majority of rockets sent into space are destined to orbit the Earth so they leave with a speed that is less than the escape speed.

Ⓗ Orbital motion

KEPLER'S THIRD LAW

The are hundreds of artificial satellites in orbit around the Earth. These satellites do not rely on any engines to keep them in orbit – the gravitational force from the Earth provides the centripetal force required.

Gravitational attraction = centripetal force

Equations for both of these quantities have been worked out elsewhere in this book

Gravitational attraction = $\frac{GMm}{r^2}$ [page 61]

Centripetal force = $\frac{m\,v^2}{r}$ [page 21]

Therefore $\frac{GMm}{r^2} = \frac{m\,v^2}{r}$ where r = radius of orbit

$$GM = v^2\,r \qquad (1)$$

$$v = \sqrt{\left(\frac{GM}{r}\right)} \qquad (2)$$

Do not confuse v (velocity of orbit) with v (escape speed).

Since the satellite does one orbit (one circumference) in time T,

$$\text{Speed } v = \frac{\text{circumference}}{T} = \frac{2\pi\,r}{T}$$

This can be substituted into equation (1) to give

$$GM = \left(\frac{2\pi\,r}{T}\right)^2 r = \frac{4\pi^2\,r^3}{T^2}$$

G, M and $(4\pi^2)$ are all constants so $\frac{r^3}{T^2}$ = constant

This is an important relationship. It is known as Kepler's third law. Although we derived it for artificial satellites in **circular** orbit around the Earth, it actually applies to ANY closed orbit.

ENERGY OF AN ORBITING SATELLITE

We already know that the gravitational energy = $\frac{-GMm}{r}$

The kinetic energy = $\frac{1}{2}\,m\,v^2$ but $v = \sqrt{\left(\frac{GM}{r}\right)}$

[equation (2), above]

$\therefore$ kinetic energy = $\frac{1}{2}\,m\,\frac{GM}{r} = \frac{1}{2}\,\frac{GMm}{r}$

So total energy = KE + PE

$$= \frac{1}{2}\,\frac{GMm}{r} - \frac{GMm}{r} = -\frac{1}{2}\,\frac{GMm}{r}$$

Note that

- In the orbit the magnitude of the KE = $\frac{1}{2}$ magnitude of the PE
- The overall energy of the satellite is negative. (A satellite must have a total energy less than zero otherwise it would have enough energy to escape the Earth's gravitional field.)

- In order to move from a small radius orbit to a large radius orbit, the total energy must increase. To be precise, an increase in orbital radius makes the total energy go from a large negative number to a smaller negative number – this is an increase.

This can be summarised in graphical form.

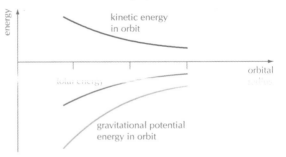

WEIGHTLESSNESS

One way of defining the weight of a person is to say that it is the value of the force recorded on a supporting scale.

If the scales were set up in a lift, they would record different values depending on the **acceleration** of the lift.

An extreme version of these situations occurs if the lift cable breaks and the lift (and passenger) accelerates down at 10 ms^{-2}.

accelerating down at 10 m s^{-2}

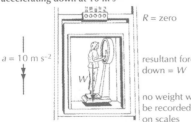

The person would appear to be weightless for the duration of the fall. Given the possible ambiguity of the term 'weight', it is better to call this situation the **apparent weightlessness** of objects in free fall together.

An astronaut in an orbiting space station would also appear weightless. The space station and the astronaut are in free fall together.

In the space station, the gravitational pull on the astronaut provides the centripetal force needed to stay in the orbit. This resultant force causes the centripetal acceleration. The same is true for the gravitational pull on the satellite and the satellite's acceleration. There is no contact force between the satellite and the astronaut so, once again, we have apparent weightlessness.

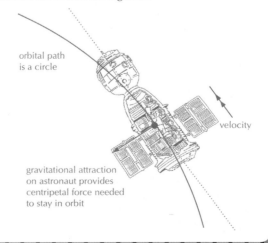

ⓗ Friction

FACTORS AFFECTING FRICTION – STATIC AND DYNAMIC

Friction is the force that opposes the relative motion of two surfaces. It arises because the surfaces involved are not perfectly smooth on the microscopic scale. If the surfaces are prevented from relative motion (they are at rest) then this is an example of **static friction**. If the surfaces are moving, then it is called **dynamic friction** or **kinetic friction**.

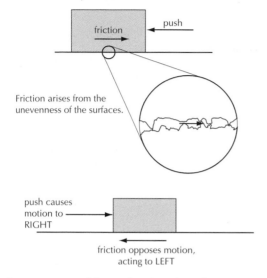

Friction arises from the unevenness of the surfaces.

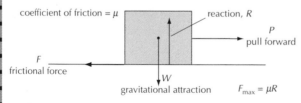

push causes motion to RIGHT

friction opposes motion, acting to LEFT

A key experimental fact is that the value of static friction changes depending on the applied force. Up to a certain maximum force, F_{max}, the resultant force is zero. For example, if we try to get a heavy block to move, any value of pushing force below F_{max} would fail to get the block to accelerate.

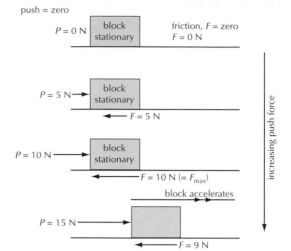

push = zero

$P = 0$ N   block stationary   friction, F = zero $F = 0$ N

$P = 5$ N → block stationary ← $F = 5$ N

$P = 10$ N → block stationary ← $F = 10$ N ($= F_{max}$)

block accelerates

$P = 15$ N → ← $F = 9$ N

increasing push force

The value of F_{max} depends upon
- the nature of the two surfaces in contact.
- the normal reaction force between the two surfaces. The maximum frictional force and the normal reaction force are proportional.

If the two surfaces are kept in contact by gravity, the value of F_{max} does NOT depend upon the area of contact

Once the object has started moving, the maximum value of friction slightly reduces. In other words,

$$F_k < F_{max}$$

For two surfaces moving over one another, the dynamic frictional force remains roughly constant even if the speed changes slightly.

COEFFICIENT OF FRICTION

Experimentally, the maximum frictional force and the normal reaction force are proportional. We use this to define the **coefficient of friction, μ**.

coefficient of friction = μ

reaction, R

P pull forward

F frictional force

W gravitational attraction $F_{max} = \mu R$

The coefficient of friction is defined from the maximum value that friction can take

$$F_{max} = \mu R$$

where R = normal reaction force

It should be noted that
- since the maximum value for dynamic friction is less than the maximum value for static friction, the values for the coefficients of friction will be different

$$\mu_k < \mu_s$$

- the coefficient of friction is a ratio between two forces – it has no units.
- if the surfaces are smooth then the maximum friction is zero i.e. $\mu = 0$.
- the coefficient of friction is less than 1 unless the surfaces are stuck together.

EXAMPLE

If a block is placed on a slope, the angle of the slope can be increased until the block just begins to slide down the slope. This turns out to be an easy experimental way to measure the coefficient of static friction.

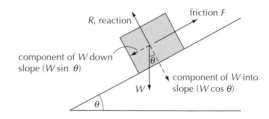

R, reaction   friction F

component of W down slope ($W \sin \theta$)

component of W into slope ($W \cos \theta$)

W

θ

If balanced,
$$F = W \sin \theta$$
$$R = W \cos \theta$$

θ is increased.
When block just starts moving,
$$F = F_{max}$$

$$\mu_{static} = \frac{F_{max}}{R}$$

$$= \frac{W \sin \theta}{W \cos \theta}$$

$$= \tan \theta$$

Static equilibrium

TURNING ACTION – TORQUES AND MOMENTS

Everyday motion tends to be a combination of **translation** and **rotation**. We can treat the two types of motions independently. A translation is when every particle in a object has the same instantaneous velocity. A rotation is when every particle in a object moves in a circle about the axis of rotation and each particle has a different velocity. A full study of rotational dynamics is outside the IB Physics diploma course.

The turning action of a force about an axis (the pivot) is called the **moment** of a force or its **torque** τ. The torque will increase if

- the value of the force is increased.
- the distance from the force to axis is increased.

It is defined as shown below

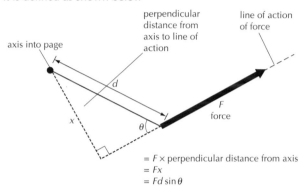

$$= F \times \text{perpendicular distance from axis}$$
$$= Fx$$
$$= Fd \sin\theta$$

Torque = force × perpendicular distance from axis of rotation
$$= Fd \sin\theta$$

If the force is perpendicular to the axis, this simplifies to

$$\tau = Fd$$

The SI unit for torque is N m. Even though this is a force multiplied by a distance it is NOT equal to a joule – work is defined as force × displacement IN THE SAME DIRECTION as the force.

EXAMPLES

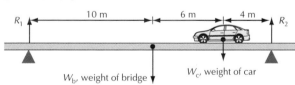

When a car goes across a bridge, the forces (on the bridge) are as shown.

Taking moments about right-hand support:
clockwise moment = anticlockwise moment
$$(R_1 \times 20 \text{ m}) = (W_b \times 10 \text{ m}) + (W_c \times 4 \text{ m})$$

Taking moments about left-hand support:
$$(R_2 \times 20 \text{ m}) = (W_b \times 10 \text{ m}) + (W_c \times 16 \text{ m})$$

Also, since bridge is not accelerating:
$$R_1 + R_2 = W_b + W_c$$

When solving problems to do with rotational equilibrium remember:

- all forces at an axis have zero moment about that axis.
- you do not have to choose the pivot as the axis about which you calculate torques, but it is often the simplest thing to do (for the reason above).
- you need to remember the sense (clockwise or anticlockwise).

ROTATIONAL EQUILIBRIUM

If an object is in rotational equilibrium, the resultant torque on it (about any axis) must be zero. This is known as the **principle of moments**.

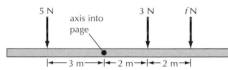

In the example above the 3 N force turns the plank clockwise about the pivot. Its torque is 6 N m clockwise. The 5 N force turns the plank anticlockwise about the pivot. Its torque is 15 N m anticlockwise. These two give a resultant torque of (15 – 6) = 9 N m anticlockwise. If the force F keeps the rod in rotational equilibrium, then it must provide a balancing torque of 9 N m clockwise. It must have a value of (9 / 4) = 2.25 N

CENTRE OF GRAVITY

The effect of gravity on all the different parts of the object can be treated as a single force acting at the object's **centre of gravity**.

(a) plank balances if pivot is in middle — centre of gravity
There is no moment about the centre of gravity.

(b) plank rotates clockwise if pivot is to the left

(c) plank rotates anticlockwise if pivot is to the right

If an object is of uniform shape and density, the centre of gravity will be in the middle of the object. If the object is not uniform, then finding its position is not trivial – it is possible for an object's centre of gravity to be outside the object. Experimentally, if you suspend an object from a point and it is free to move, then the centre of gravity will always end up below the point of suspension.

- when solving two-dimensional problems it is sufficient to show that an object is in rotational equilibrium about any ONE axis.
- Newton's laws still apply. Often an object is in rotational AND in translation equilibrium. This can provide a simple way of finding an unknown force.
- the weight of an object can be considered to be concentrated at its centre of gravity.
- if the problem only involves three non-parallel forces, the lines of action of all the forces must meet at single point in order to be in rotational equilibrium.

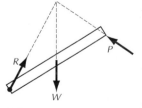

3 forces must meet at a point if in equilibrium

1 A force P directed up a plane, of angle θ to the horizontal, prevents a block of mass M from slipping down, as shown in the diagram.

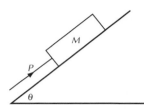

If the coefficient of static friction between the block and the plane is μ, the **minimum** magnitude of force P required is

A $Mg \sin \theta$ **C** $Mg (\sin \theta + \mu \cos \theta)$

B $Mg (\sin \theta - \mu \cos \theta)$ **D** $Mg (\cos \theta - \mu \sin \theta)$

2 The Earth has approximately 81 times the mass of the Moon. There is a point between the Earth and the Moon where their resultant gravitational field is zero. if the distance to this point from the centre of the Earth is y and from the centre of the Moon it is x, the ratio of $\frac{y}{x}$ is approximately

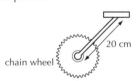

A $(81)^{1/4}$ **B** $(81)^{1/2}$ **C** 81 **D** 81^2

3 A particle of mass m moves with constant speed v in a circle of radius r. The work done on the particle by the centripetal force in one complete revolution is

A $2\pi m v^2$ **B** $\frac{2\pi v^2}{m}$ **C** $\frac{2\pi m}{v^2}$ **D** zero

4 A 60 kg person riding a bicycle puts all of her weight on each pedal in turn when climbing a hill. The pedals rotate in a circle of radius 20 cm. Estimate the maximum torque that is exerted by the person.

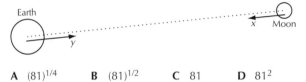

A 12 Nm **C** 1200 Nm

B 120 Nm **D** 12000 Nm

5 A projectile is launched **horizontally** from a high tower. Which **one** of the following graphs best represents the **vertical component** of the projectile's velocity from the time it is launched to the time it hits the ground? Assume negligible air resistance.

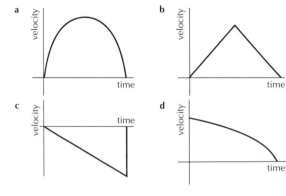

6 The Space Shuttle orbits about 300 km above the surface of the Earth. The shape of the orbit is circular, and the mass of the Space Shuttle is 6.8×10^4 kg. The mass of the Earth is 6.0×10^{24} kg, and radius of the Earth is 6.4×10^6 m.

(a) **(i)** Calculate the change in the Space Shuttle's gravitational potential energy between its launch and its arrival in orbit. [3]

(ii) Calculate the speed of the Space Shuttle whilst in orbit. [2]

(iii) Calculate the energy needed to put the Space Shuttle into orbit. [2]

(b) **(i)** What forces, if any, act on the astronauts inside the Space Shuttle whilst in orbit? [1]

(ii) Explain why astronauts aboard the Space Shuttle feel weightless. [2]

(c) Imagine an astronaut 2 m outside the exterior walls of the Space Shuttle, and 10 m from the centre of mass of the Space Shuttle. By making appropriate assumptions and approximations, calculate how long it would take for this astronaut to be pulled back to the Space Shuttle by the force of gravity alone. [7]

7 **(a)** The diagram below shows a planet of mass M and radius R_p.

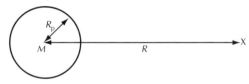

The gravitational potential V due to the planet at point X distance R from the centre of the planet is given by

$$V = - \frac{GM}{R}$$

where G is the universal gravitational constant.

Show that the gravitational potential V can be expressed as

$$V = - \frac{g_0 R_p^2}{R}$$

where g_0 is the acceleration of free fall at the surface of the planet. [3]

(b) The graph below shows how the gravitational potential V due to the planet varies with distance R from the centre of the planet for values of R greater than R_p, where $R_p = 2.5 \times 10^6$ m.

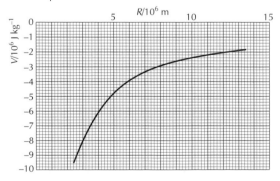

Use the data from the graph to

(i) determine a value of g_0. [2]

(ii) show that the minimum energy required to raise a satellite of mass 3000 kg to a height 3.0×10^6 m above the **surface** of the planet is about 1.7×10^{10} J. [3]

HL Thermodynamic systems and concepts

DEFINITIONS

Historically, the study of the behaviour of ideal gases led to some very fundamental concepts that are applicable to many other situations. These laws, otherwise known as the laws of **thermodynamics,** provide the modern physicist with a set of very powerful intellectual tools.

The terms used need to be explained.

Thermodynamic system	Most of the time when studying the behaviour of an ideal gas in particular situations, we focus on the macroscopic behaviour of the gas as a whole. In terms of work and energy, the gas can gain or lose thermal energy and it can do work or work can be done on it. In this context, the gas can be seen as a **thermodynamic system**.
The surroundings	If we are focusing our study on the behaviour of an ideal gas, then everything else can be called its **surroundings**. For example the expansion of a gas means that work is done by the gas on the surroundings (see below).
Heat ΔQ	In this context heat refers to the transfer of a quantity of thermal energy between the system and its surroundings. This transfer must be as a result of a temperature difference. 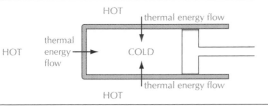
Work ΔW	In this context, work refers to the macroscopic transfer of energy. For example 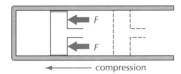 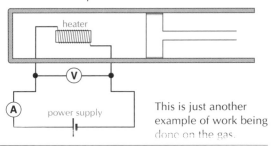

1. work done = force × distance

When a gas is compressed, work is done on the gas

When a gas is compressed, the surroundings do work on it. When a gas expands it does work on the surroundings.

2. work done = potential difference × current × time

This is just another example of work being done on the gas.

Internal energy U ΔU = *change in internal energy*	The internal energy can be thought of as the energy held within a system. It is the sum of the PE due to the intermolecular forces and the kinetic energy due to the random motion of the molecules. See page 24.

This is different to the total energy of the system, which would also include the overall motion of the system and any PE due to external forces.

In thermodynamics, it is the changes in internal energy that are being considered. If the internal energy of a gas is increased, then its temperature must increase. A change of phase (e.g. liquid → gas) also involves a change of internal energy.

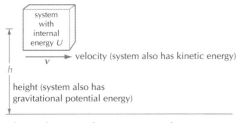

The total energy of a system is not the same as its internal energy

WORK DONE DURING EXPANSION AT CONSTANT PRESSURE

Whenever a gas expands, it is doing work on its surroundings. If the pressure of the gas is changing all the time, then calculating the amount of work done is complex. This is because we cannot assume a constant force in the equation of work done (work done = force × distance). If the pressure changes then the force must also change. If the pressure is constant then the force is constant and we can calculate the work done.

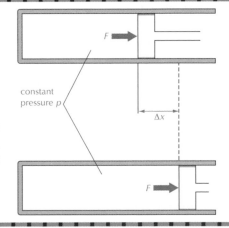

Work done ΔW = force × distance

$$= F \Delta x$$

Since pressure = $\dfrac{\text{force}}{\text{area}}$

$$F = pA$$

therefore

$$\Delta W = pA\Delta x$$

but $A\Delta x = \Delta V$

so work done $= p \, \Delta V$

So if a gas increases its volume (ΔV is positive) then the gas does work (ΔW is positive)

FIRST LAW OF THERMODYNAMICS

There are three fundamental laws of thermodynamics. The first law is simply a statement of the principle of energy conservation as applied to the system. If an amount of thermal energy ΔQ is given to a system, then one of two things must happen (or a combination of both). The system can increase its internal energy ΔU or it can do work ΔW.

As energy is conserved

$$\Delta Q = \Delta U + \Delta W$$

It is important to remember what the signs of these symbols mean. They are all taken from the system's 'point of view'.

ΔQ If this is **positive**, then thermal energy is going into the system.
If it is **negative**, the thermal energy is going out of the system.

ΔU If this is **positive**, then the internal energy of the system is **increasing**.
(The temperature of the gas is increasing.)
If it is **negative**, the internal energy of the system is **decreasing**.
(The temperature of the gas is decreasing.)

ΔW If this is **positive**, then the **system is doing work** on the surroundings.
(The gas is expanding.)
If it is **negative**, the **surroundings are doing work** on the system.
(The gas is contracting.)

p-V DIAGRAMS AND WORK DONE

It is often useful to represent the changes that happen to a gas during a thermodynamic process on a p-V diagram. An important reason for choosing to do this is that the area under the graph represents the work done. The reasons for this are shown below.

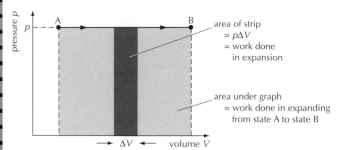

area of strip
= $p\Delta V$
= work done in expansion

area under graph
= work done in expanding from state A to state B

This turns out to be generally true for any thermodynamic process.

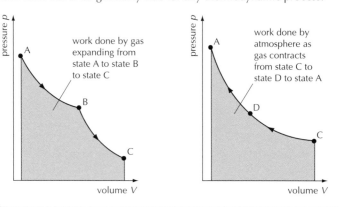

work done by gas expanding from state A to state B to state C

work done by atmosphere as gas contracts from state C to state D to state A

IDEAL GAS PROCESSES

A gas can undergo any number of different types of change or process. Four important process are considered below. In each case the changes can be represented on a pressure – volume diagram and the first law of thermodynamics must apply. To be precise, these diagrams represent a type of process called a reversible process.

1. Isochoric (isovolumetric)

In an isochoric process, also called an isovolumetric process, the gas has a constant volume. The diagram below shows an **isochoric decrease** in pressure.

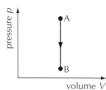

Isochoric (volumetric) change

$V = constant$, or $\dfrac{p}{T} = constant$

ΔQ negative
ΔU negative
ΔW zero

2. Isobaric

In an isobaric process the gas has a constant pressure. The diagram below shows an **isobaric expansion**.

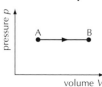

Isobaric change

$p = constant$, or $\dfrac{V}{T} = constant$

ΔQ positive
ΔU positive
ΔW positive

3. Isothermal

In an isothermal process the gas has a constant temperature. The diagram below shows an **isothermal expansion**.

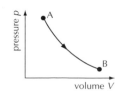

Isothermic change

$T = constant$, or $pV = constant$

ΔQ positive
ΔU zero
ΔW positive

4. Adiabatic

In an adiabatic process there is no thermal energy transfer between the gas and the surrounds. This means that if the gas does work it must result in a decrease in internal energy. A rapid compression or expansion is approximately adiabatic. This is because done quickly there is not sufficient time for thermal energy to be exchanged with the surroundings. The diagram below shows an **adiabatic expansion**.

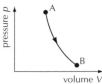

Adiabatic change

ΔQ zero
ΔU negative
ΔW positive

Heat engines and heat pumps

HEAT ENGINES

A central concept in the study of thermodynamics is the **heat engine**. A heat engine is any device that uses a source of thermal energy in order to do work. It converts heat into work. The internal combustion engine in a car and the turbines that are used to generate electrical energy in a power station are both examples of heat engines. A block diagram representing a generalised heat engine is shown below.

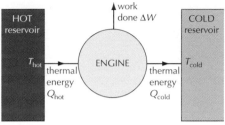

Heat engine

In this context, the word **reservoir** is used to imply a constant temperature source (or sink) of thermal energy. Thermal energy can be taken from the hot reservoir without causing the temperature of the hot reservoir to change. Similarly thermal energy can be given to the cold reservoir without increasing its temperature.

An ideal gas can be used as a heat engine. The p-V diagram right represents a simple example. The four-stage cycle returns the gas to its starting conditions, but the gas has done work. The area enclosed by the cycle represents the amount of work done.

In order to do this, some thermal energy must have been taken from a hot reservoir (during the isovolumetric increase in pressure and the isobaric expansion). A different amount of thermal energy must have been ejected to a cold reservoir (during the isovolumetric decrease in pressure and the isobaric compression).

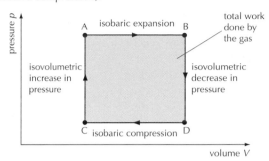

The thermal efficiency of a heat engine is defined as

$$\text{Efficiency} = \frac{\text{work done}}{\text{(thermal energy taken from hot reservoir)}}$$

This is equivalent to

$$\text{Efficiency} = \frac{\text{rate of doing work}}{\text{(thermal power taken from hot reservoir)}}$$

In symbols, efficiency $= \frac{\Delta W}{Q_{\text{hot}}}$

The cycle of changes that results in a heat engine with the maximum possible efficiency is called the **Carnot cycle**.

HEAT PUMPS

A **heat pump** is a heat engine being run in reverse. A heat pump causes thermal energy to be moved from a cold reservoir to a hot reservoir. In order for this to be achieved, mechanical work must be done.

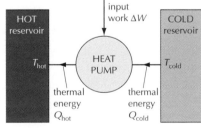

Heat pump

Once again an ideal gas can be used as a heat pump. The thermodynamic processes can be exactly the same ones as were used in the heat engine, but the processes are all opposite. This time an anticlockwise circuit will represent the cycle of processes.

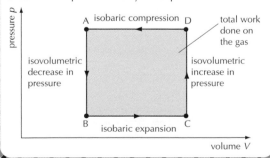

CARNOT CYCLES AND CARNOT THEOREM

The Carnot cycle represents the cycle of processes for a theoretical heat engine with the maximum possible efficiency. Such an idealised engine is called a **Carnot engine**.

It consists of an ideal gas undergoing the following processes.

- Isothermal expansion (A → B)
- Adiabatic expansion (B → C)
- Isothermal compression (C → D)
- Adiabatic compression (D → A)

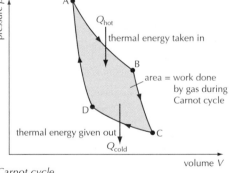

Carnot cycle

The temperatures of the hot and cold reservoirs fix the maximum possible efficiency that can be achieved.

The efficiency of a Carnot engine can be shown to be

$$e_c = 1 - \left(\frac{T_{\text{cold}}}{T_{\text{hot}}} \right) \text{ (where } T \text{ is in kelvin)}$$

An engine operates at 300°C and ejects heat to the surroundings at 20°C. The maximum possible theoretical efficiency

$= 1 - \frac{293}{573}$

$= 0.49 \quad = 49\%$

HL Second law of thermodynamics and entropy

SECOND LAW OF THERMODYNAMICS

Historically the **second law of thermodynamics** has been stated in many different ways. All of these versions can be shown to be equivalent to one another.

In principle there is nothing to stop the complete conversion of thermal energy into useful work. In practice, a gas can not continue to expand forever – the apparatus sets a physical limit. Thus **the continuous conversion of thermal energy into work requires a cyclical process – a heat engine.**

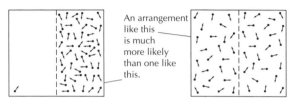

Carnot showed that $Q_{hot} > \Delta W$.

In other words there must be thermal energy 'wasted' to the cold reservoir.

This realisation leads to possibly the simplest formulation of the second law of thermodynamics (the **Kelvin-Planck** formulation).

> **No heat engine, operating in a cycle, can take in heat from its surroundings and totally convert it into work**

Other possible formulations include the following:

> **No heat pump can transfer thermal energy from a low-temperature reservoir to a high-temperature reservoir without work being done on it** (Clausius)

> **Heat flows from hot objects to cold objects**

The concept of **entropy** leads to one final version of the second law.

> **The entropy of the universe can never decrease**

EXAMPLES

The first and second laws of thermodynamics both must apply to all situations. Local decreases of entropy are possible so long as elsewhere, there is a corresponding increase.

1. A refrigerator is an example of a heat pump.

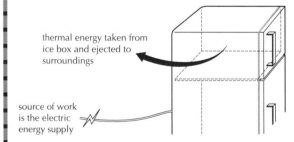

thermal energy taken from ice box and ejected to surroundings

source of work is the electric energy supply

A refrigerator

2. It should be possible to design a theoretical system for propelling a boat based around a heat engine. The atmosphere could be used as the hot reservoir and cold water from the sea could be used as the cold reservoir. The movement of the boat through the water would be the work done. This is possible BUT it cannot continue to work for ever. The sea would be warmed and the atmosphere would be cooled and eventually there would be no temperature difference.

ENTROPY AND ENERGY DEGRADATION

Entropy is a property that expresses the disorder in the system.

The details are not important but the entropy S of a system is linked to the number of possible arrangements LW of the system.

Because molecules are in random motion, one would expect roughly equal numbers of gas molecules in each side of a container.

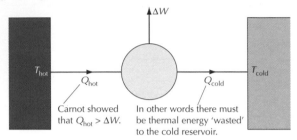

An arrangement like this is much more likely than one like this.

The number of ways of arranging the molecules to get the set-up on the right is greater than the number of ways of arranging the molecules to get the set-up on the left. This means that the entropy of the system on the right is greater than the entropy of the system on the left.

In any random process the amount of disorder will tend to increase. In other words, the total entropy will always increase. The entropy change ΔS is linked to the thermal energy change ΔQ and the temperature T. ($\Delta S = \Delta Q / T$.)

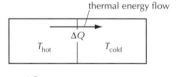

thermal energy flow

ΔQ

T_{hot} T_{cold}

$$\text{decrease of entropy} = \frac{\Delta Q}{T_{hot}} \qquad \text{increase of entropy} = \frac{\Delta Q}{T_{cold}}$$

When thermal energy flows from a hot object to a colder object, overall the total entropy has increased.

In many situations the idea of energy **degradation** is a useful concept. The more energy is shared out, the more degraded it becomes – it is harder to put it to use. For example, the internal energy that is 'locked' up in oil can be released when the oil is burned. In the end, all the energy released will be in the form of thermal energy – shared amongst many molecules. It is not feasible to get it back.

3. Water freezes at 0 °C because this is the temperature at which the entropy increase of the surroundings (when receiving the latent heat) equals the entropy decrease of the water molecules becoming more ordered. It would not freeze at a higher temperature because this would mean that the overall entropy of the system would decrease.

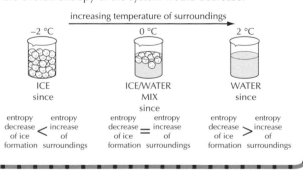

increasing temperature of surroundings

–2 °C	0 °C	2 °C
ICE	ICE/WATER MIX	WATER
since	since	since

| entropy decrease of ice formation | < | entropy increase of surroundings | entropy decrease of ice formation | = | entropy increase of surroundings | entropy decrease of ice formation | > | entropy increase of surroundings |

1 An ideal gas expands **isothermically**, absorbing a certain amount of energy, Q, in the process. It then returns to its original volume **adiabatically**. During the adiabatic process, the internal energy change of gas will be

A   zero.

B   smaller than Q.

C   equal to Q.

D   greater than Q.

2 When the volume of a gas increases it does work. The work done would be greatest for which one of the following processes?

A   Isobaric.

B   Adiabatic.

C   Isothermal.

D   The work done would be the same for all of the above.

3 It is proposed to build a heat engine that would operate between a hot reservoir at a temperature of 400K and a cold reservoir at 300 K. See the diagram below. In each cycle it would take 100 J from the hot reservoir, lose 25 J to the cold reservoir and do 75 J of work.

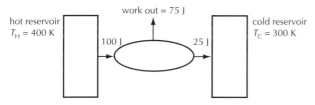

This proposed heat engine would violate

A   both the first and the second laws of theormodynamics.

B   the first but not the second law of thermodynamics.

C   the second but not the first law of thermodynamics.

D   neither the first nor the second law of thermodynamics.

4 The energy absorbed by an ideal gas during an isothermal expansion is equal to

A   the work done by the gas.

B   the work done on the gas.

C   the change in the internal energy of the gas.

D   zero.

5 A fixed mass of a gas undergoes various changes of temperature, pressure and volume such that it is taken round the p–V cycle shown in the diagram below.

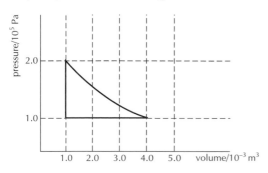

The following sequence of processes takes place during the cycle.

X → Y   the gas expands at constant temperature and the gas absorbs energy from a reservoir and does 450 J of work.

Y → Z   the gas is compressed and 800 J of thermal energy is transferred from the gas to a reservoir.

Z → X   the gas returns to its initial stage by absorbing energy from a reservoir.

(a) Is there a change in internal energy of the gas during the processes X → Y? Explain. [2]

(b) Is the energy absorbed by the gas during the process X → Y less than, equal to or more than 450 J? Explain. [2]

(c) Use the graph to determine the work done on the gas during the process Y → Z. [3]

(d) What is the change in internal energy of the gas during the process Y → Z? [2]

(e) How much thermal energy is absorbed by the gas during the process Z → X? Explain your answer. [2]

(f) What quantity is represented by the area enclosed by the graph? Estimate its value. [2]

(g) The overall efficiency of a heat engine is defined as

$$\text{Efficiency} = \frac{\text{net work done by the gas during a cycle}}{\text{total energy absorbed during a cycle}}$$

If this p–V cycle represents the cycle for a particular heat engine determine the efficiency of the heat engine. [2]

6 In a **diesel** engine, air is initially at a pressure of 1×10^5 Pa and a temperature of 27 °C. The air undergoes the cycle of changes listed below. At the end of the cycle, the air is back at its starting conditions.

1   An **adiabatic compression** to 1/20th of its original volume.

2   A brief **isobaric expansion** to 1/10th of its original volume.

3   An **adiabatic expansion** back to its original volume.

4   A cooling down at constant volume.

(a) Use the axes below to sketch, with labels, the cycle of changes that the gas undergoes. Accurate values are not required. [3]

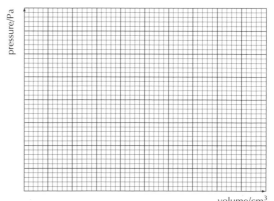

(b) If the pressure after the **adiabatic compression** has risen to 6.6×10^6 Pa, calculate the temperature of the gas. [2]

(c) In which of the four processes:
   (i)   is work done **on** the gas? [1]
   (ii) is work done **by** the gas? [1]
   (iii) does ignition of the air-fuel mixture take place? [1]

(d) Explain how the 2nd law of thermodynamics applies to this cycle of changes. [2]

The Doppler effect

MATHEMATICS OF THE DOPPLER EFFECT

Mathematical equations that apply to sound are derived below.

Unfortunately the same analysis does not apply to light – the velocities can not be worked out relative to the medium. It is, however, possible to derive an equation for light that turns out to be in exactly the same form as the equation for sound as long as two conditions are met:
- the relative velocity of source and detector is used in the equations.
- this relative velocity is a lot less than the speed of light.

MOVING SOURCE

Source moves from A to D.

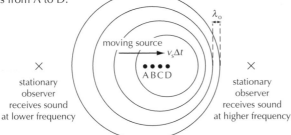

in a time Δt:
$f_s\Delta t$ waves are compressed into $(c - v_s)\Delta t$
(where c = speed of sound)

$\therefore$ received wavelength, $\lambda_o = \dfrac{\text{distance}}{\text{number of waves}}$

$= \dfrac{(c - v_s)\Delta t}{f_s \Delta t}$

$= \dfrac{c - v_s}{f_s}$

Received frequency

$f_o = \dfrac{c}{\lambda_o}$

$= \dfrac{f_s}{c - v_s} \times c$

$f_o = f_s\left(\dfrac{1}{1 - \frac{v_s}{c}}\right)$

If the source is going away:

$f_o = f_s\left(\dfrac{1}{1 + \frac{v_s}{c}}\right)$

MOVING DETECTOR

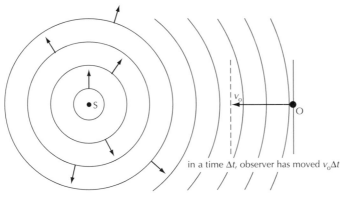

in a time Δt, observer has moved $v_o\Delta t$

In a time Δt:

the approaching observer receives an additional number of waves.

$\therefore \quad f_o = f_s + \left(\dfrac{\text{extra waves in } v_o\Delta t}{\Delta t}\right)$

$= f_s + \dfrac{v_o\Delta t}{\lambda \Delta t}$

$= f_s + \dfrac{v_o}{\left(\frac{c}{f_s}\right)}$

$= f_s + f_s \times \dfrac{v_o}{c}$

$f_o = f_s\left(1 + \dfrac{v_o}{c}\right)$

If observer is moving away:

$f_o = f_s\left(1 - \dfrac{v_o}{c}\right)$

EXAMPLE

The frequency of a car's horn is measured by a stationary observer as 200Hz when the car is at rest. What frequency will be heard if the car is approaching the observer at 30 m s⁻¹? (Speed of sound in air is 330 m s⁻¹)

$f_s = 200$ Hz

$f_o = ?$

$v_s = 30$ m s⁻¹

$c = 330$ m s⁻¹

$f_o = 200\left(\dfrac{1}{1 - \frac{30}{300}}\right)$

$= 200 \times 1.1$

$= 220$ Hz

⬤HL Beats

Beats

BEATS

On page 34 we saw how the principle of superposition can be applied whenever waves meet. This includes sound waves. **Beats** are formed when two sound waves interfere that are:

- of similar **but not identical** frequency
- of similar or equal amplitude

They are called beats because the sound that you hear is a single frequency sound but one that varies in amplitude. That is, it varies in loudness (loud then soft then loud then soft etc.). Typically this cycle happens several times every second. Musicians tune their instruments using these beats. As an instrument is brought into tune, the beat frequency gets less and less. If two instruments were producing exactly the same frequency note, there would be no beats.

In order to explain why these beats occur, we simply apply the principle of superposition and work out the resultant waveform from the individual displacements due to each wave.

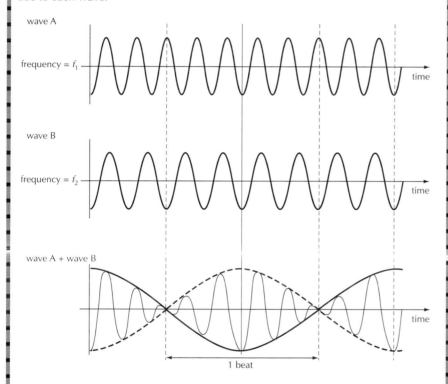

If the two frequencies that are creating the beats are f_1 and f_2, then

- the frequency of the sound that is heard is equal to the average frequency, $\left(\dfrac{f_1 + f_2}{2}\right)$. The frequencies are normally so close to one another that the frequency heard can be taken to be roughly equal to either frequency.
- the beat frequency is equal to the difference between the two frequencies, $f_{beat} = f_1 - f_2$

These formulae can be derived from the graphs. One full beat (loud to soft and back to loud) takes place when the waves move from being in phase to being out of phase and return back to being in phase. In the time it takes for this to happen (the beat period), one wave has achieved one full cycle less than the other. In one whole second, one wave completes f_1 cycles whereas the other wave will complete f_2 cycles. The number of beats in one second will be the difference between these two numbers, $f_1 - f_2$. Thus the beat frequency will be $f_1 - f_2$.

EXAMPLE

Beats are not confined to sound waves – in principle they can happen whenever two waves of similar, but not equal, frequencies meet. An interesting example is the formation of tides on the Earth. The Sun causes two tides every day (twenty-four hours). The moon causes two tides every twenty-five hours. These two similar frequencies form beats. When they are in phase, the tides are very large (spring tides) when they are out of phase, the tides are small (neap tides).

Two-source interference of waves

PRINCIPLES OF THE TWO-SOURCE INTERFERENCE PATTERN

Two-source interference is simply another application of the principle of superposition, for two coherent sources have roughly the same amplitude.

Two sources are coherent if
* they have the same frequency.
* there is a constant phase relationship between the two sources.

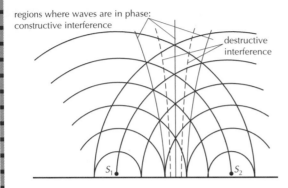

regions where waves are in phase: constructive interference

destructive interference

S_1 S_2

Two loudspeakers both connected to the same signal generator are coherent sources. This forms regions of loud and soft sound.

A set-up for viewing two-source interference with light is shown below. It is known as **Young's double slit** experiment. A **monochromatic** source of light is one that gives out only one frequency. Light from the twin slits (the sources) interferes and patterns of light and dark are dark regions, called **fringes**, can be seen on the screen.

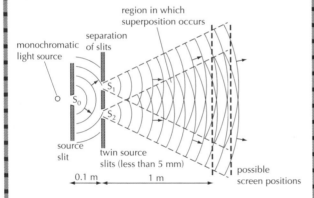

region in which superposition occurs

monochromatic light source

separation of slits

S_0 S_1 S_2

source slit

twin source slits (less than 5 mm)

0.1 m   1 m

possible screen positions

The use of a laser makes the set-up easier.

laser

double slit

screen

The experiment results in a regular pattern of light and dark strips across the screen as represented below. See page 170.

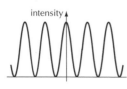 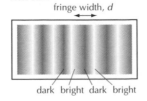

intensity distribution

view seen

fringe width, d

intensity

dark bright dark bright

MATHEMATICS

The location of the light and dark fringes can be mathematically derived in one of two ways.

Method 1

The simplest way is to consider two parallel rays setting off from the slits as shown below.

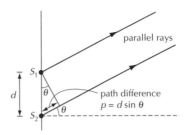

parallel rays

S_1

d

θ

S_2

θ

path difference $p = d \sin \theta$

If these two rays result in a bright patch, then the two rays must arrive in phase. The two rays of light started out in phase but the light from source 2 travels an extra distance. This extra distance is called the **path difference**.

Constructive interference can only happen if the path difference is a whole number of wavelengths. Mathematically,

Path difference = $n \lambda$

[where n is an integer – e.g. 1, 2, 3 etc.]

From the geometry of the situation

Path difference = $d \sin \theta$

In other words $n \lambda = d \sin \theta$

Method 2

If a screen is used to make the fringes visible, then the rays from the two slits cannot be absolutely parallel, but the physical set-up means that this is effectively true.

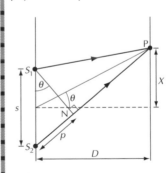

S_1 θ θ P

X

s N

S_2 p

D

$\sin \theta = \dfrac{p}{s}$

$\tan \theta = \dfrac{X}{D}$

If θ is small $\sin \theta = \tan \theta$

so $\dfrac{p}{s} = \dfrac{X}{D}$

$\therefore$ $p = \dfrac{Xs}{D}$

For constructive interference:

$p = n\lambda$

$\therefore$ $n\lambda = \dfrac{X_n s}{D}$

$\therefore$ $X_n = \dfrac{n \lambda D}{s}$

fringe width, $d = X_{n+1} - X_n$

$= \dfrac{\lambda D}{s}$

$\therefore$ $s = \dfrac{\lambda D}{d}$

Example

Laser light of wavelength 450 mm is shone on two slits that are 0.1 mm apart. How far apart are the fringes on a screen placed 5.0 m away?

$$d = \frac{\lambda D}{s} = \frac{4.5 \times 10^{-7} \times 5}{1.0 \times 10^{-4}} = 0.0225 \text{ m} = 2.25 \text{ cm}$$

1 Listed below are frequencies of tuning forks that are to be sounded in pairs. The largest number of beats will be heard from which pair of tuning forks?

A 201 and 200 Hz   **C** 535 and 540 Hz

B 253 Hz and 260 Hz   **D** 1420 and 1424 Hz

2 Light from two sources can produce interference fringes on a screen provided that the two sources are coherent. The word **coherent** means that the two sources

A have the same amplitude.

B need to be closer together.

C have a fixed phase relationship.

D emit polarised light.

3 A train approaches, and passes through, a station. During this period the velocity of the train is constant and the engine is continuously sounding its whistle. Which one of the following correctly describes what an observer on the platform will hear?

	Sound heard as the train is **approaching** the station	Sound heard as the train is **passing through** the station
A	Constant frequency	Increasing frequency
B	Increasing frequency	Decreasing frequency
C	Decreasing frequency	Increasing frequency
D	Constant frequency	Decreasing frequency

4 Light of wavelength λ is incident on two slits and produces a pattern of bright and dark fringes on a screen beyond the slits, as shown.

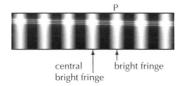

central bright fringe   bright fringe

For the bright fringe marked P, what is the difference in path length from P to the two slits, in terms of the wavelength λ?

A zero   **B** $\frac{\lambda}{2}$ **C** λ **D** 2λ

5 When a train travels towards you sounding its whistle, the pitch of the sound you hear is different from when the train is at rest. This is because

A the sound waves are travelling faster toward you.

B the wavefronts of the sound reaching you are spaced closer together.

C the wavefronts of the sound reaching you are spaced further apart.

D the sound frequency emitted by the whistle changes with the speed of the train.

6 Two tuning forks of frequencies 250 Hz and 254 Hz are sounded together. Which **one** of the following will be heard?

A Two distinguishable notes, one at 250 Hz and the other at 254 Hz.

B A note at 252 Hz of constant loudness.

C A note at 252 Hz fluctuating in loudness.

D A note which alternates in pitch between 250 and 254 Hz.

7 A sound source emits a note of constant frequency. An observer is travelling in a straight line towards the source at a constant speed. As she approaches the source she will hear a sound that

A gets higher and higher in frequency.

B gets lower and lower in frequency.

C is of constant frequency but of a frequency higher than that of the sound from the source.

D is of constant frequency but of a frequency lower than that of the sound from the source.

8 A car is travelling at constant speed towards a stationary observer whilst its horn is sounded. The frequency of the note emitted by the horn is 660 Hz. The observer, however, hears a note of frequency 720 Hz.

(a) With the aid of a diagram, explain why a higher frequency is heard. [2]

(b) If the speed of sound is 330 m s^{-1}, calculate the speed of the car. [2]

(c) This situation does not involve a **shock wave**. With the aid of a diagram, explain what is meant by a shock wave. [2]

9 This question is about the formation of beats in sound waves.

(a) State the principle of linear superposition as applied to waves. [2]

Two tuning forks **A** and **B** of slightly different frequencies are sounded simultaneously, producing two sound waves of the same amplitude. The figure below shows the disturbance at a particular point in the air as a function of time for each of the tuning forks separately, and the resultant disturbance **C**.

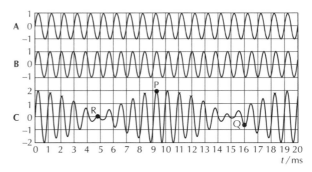

(b) Three points on the resultant waveform are labelled **P**, **Q** and **R**. For **each** of these points, check whether the resultant waveform **C** as drawn is correct, by referring to the two component waves. Explain in each case. [3]

(c) Use the diagram to determine
  (i) the frequencies of **A** and **B**; [2]
  (ii) the beat frequency. [2]

(d) (i) Beats at this frequency could not actually be perceived as beats by human hearing. Explain why. [1]
  (ii) In order that the beats become perceived as such by the ear, would the difference in frequency between **A** and **B** have to be greater or smaller than in the case above? [1]

(e) Explain how use could be made of beats to tune a guitar string against a tuning fork. [2]

Electric potential due to a point charge

POTENTIAL AND POTENTIAL DIFFERENCE

The concept of electrical potential difference between two points was introduced on page 41. As the name implies, potential difference is just the difference between the potential at one point and the potential at another. Potential is simply a measure of the total electrical energy per unit charge at a given point in space. The definition is very similar to that of gravitational potential.

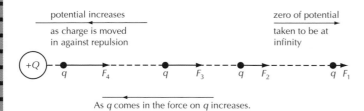

As q comes in the force on q increases.

If the total work done in bringing a positive test charge q from infinity to a point in an electric field is W, then the electric potential at that point, V, is defined to be

$$V = \frac{W}{q}$$

The units for potential are the same as the units for potential difference: $J\,C^{-1}$ or volts.

$$V = \frac{Q}{4\pi\varepsilon_0 r}$$

This equation only applies to a single point charge.

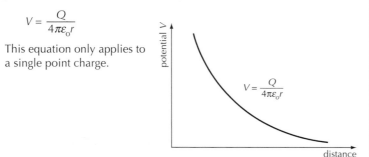

POTENTIAL DUE TO MORE THAN ONE CHARGE

If several charges all contribute to the total potential at a point, it can be calculated by adding up the individual potentials due to the individual charges.

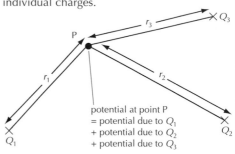

potential at point P
= potential due to Q_1
+ potential due to Q_2
+ potential due to Q_3

The electric potential at any point outside a charged conducting sphere is exactly the same as if all the charge had been concentrated at its centre.

POTENTIAL AND FIELD STRENGTH

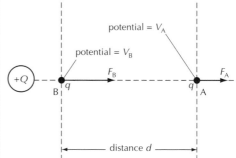

Bringing a positive charge from A to B means work needs to be done against the electrostatic force.

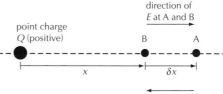

direction of force applied by external agent on test charge q_0 at A and B

The work done $\delta W = -E q\,\delta x$ [the negative sign is because the direction of the force needed to do the work is opposite to the direction of E]

Therefore $\quad E = -\dfrac{1}{q}\dfrac{\delta W}{\delta x}$

$$= -\frac{\delta V}{\delta x} \quad [\text{since } \delta V = \frac{\delta W}{q}]$$

In words,

electric field = – potential gradient

$$\text{Units} = \frac{\text{volt}}{\text{metre}} \; (Vm^{-1})$$

COMPARISON WITH GRAVITY

Electrostatics	Gravitational
Coulomb's law – for point charges	Newton's law – for point masses
$F = \dfrac{q_1 q_2}{4\pi\varepsilon_0 r^2}$	$F = G\dfrac{m_1 m_2}{r^2}$
Electric field	Gravitational field
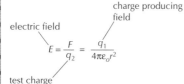	$g = \dfrac{F}{m_2} = \dfrac{Gm_1}{r^2}$
Electric potential due to a point charge	Gravitational potential due to a point mass
$V_{\text{elec}} = \dfrac{q_1}{4\pi\varepsilon_0 r}$	$V_{\text{grav}} = -\dfrac{Gm_1}{r}$
Force can be attractive or repulsive	Force always attractive

HL Equipotentials

EQUIPOTENTIAL SURFACES

The best way of representing how the electric potential varies around a charged object is to identify the regions where the potential is the same. These are called **equipotential** surfaces. In two dimensions they would be represented as lines of equipotential. A good way of visualising these lines is to start with the contour lines on a map.

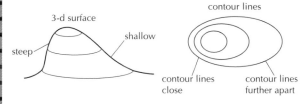

The contour diagram on the right represents the changing heights of the landscape on the left. Each line joins up points that are at the same height. Points that are high up represent a high value of gravitational potential and points that are low down represent a low gravitational potential. Contour lines are lines of equipotential in a gravitational field.

The same can be done with an electric field. Lines are drawn joining up points that have the same electric potential. The situation below shows the equipotentials for a isolated positive point charge.

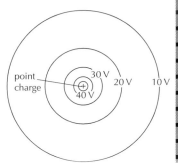

RELATIONSHIP TO FIELD LINES

There is a simple relationship between electric field lines and lines of equipotential – they are always at right angles to one another. Imagine the contour lines. If we move along a contour line, we stay at the same height in the gravitational field. This does not require work because we are moving at right angles to the gravitational force. Whenever we move along an electric equipotential line, we are moving between points that have the same electric potential – in other words, no work is being done. Moving at right angles to the electric field is the only way to avoid doing work in an electric field. Thus equipotential lines must be at right angles to field lines as shown below.

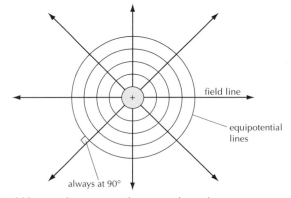

Field lines and equipotentials are at right angles

EXAMPLES OF EQUIPOTENTIALS

The diagrams below show equipotential lines for various situations.

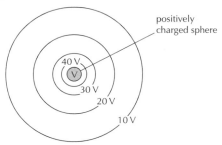

Equipotentials outside a charge-conducting sphere are the same as those for a point charge.

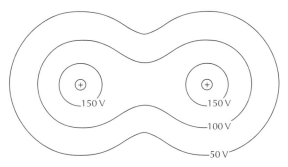

Equipotentials for two point charges (same charge)

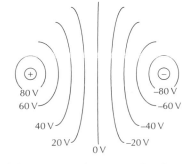

Equipotentials for two point charges (equal and opposite charges)

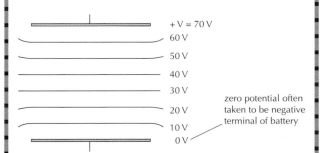

Equipotentials lines between charged parallel plates

It should be noted that although the correct definition of zero potential is at infinity, most of the time we are not really interested in the actual value of potential, we are only interested in the value of the difference in potential. This means that in some situations (such as the parallel conducting plates) it is easier to imagine the zero at a different point. This is just like setting sea level as the zero for gravitational contour lines rather than correctly using infinity for the zero.

PRODUCTION OF INDUCED E.M.F. BY RELATIVE MOTION

An e.m.f. is induced in a conductor whenever flux is cut. But flux is more than just a way of picturing the situation.

If the magnetic field is perpendicular to the surface, the magnetic flux $\Delta\phi$ passing through the area ΔA is defined in terms of the magnetic field strength B as follows.

$$\Delta\phi = B\,\Delta A, \text{ so } B = \frac{\Delta\phi}{\Delta A}$$

In a uniform field, $B = \dfrac{\phi}{A}$

An alternative name for 'magnetic field strength' is '**flux density**'.

If the area is not perpendicular, but at an angle θ to the field lines, the equation becomes

$$\phi = B\,A\,\cos\theta \text{ (units: T m}^{-2})$$

Flux can also be measured in webers (Wb), defined as follows.

$$1\text{ Wb} = 1\text{T m}^{-2}$$

This definition allows us to calculate the induced e.m.f. in terms of flux.

in a time Δt:

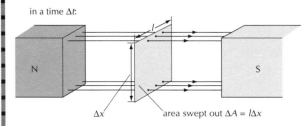

area swept out $\Delta A = l\Delta x$

$\varepsilon = B\,l\,v$ since $v = \dfrac{\Delta x}{\Delta t}$ then $\varepsilon = \dfrac{B\,l\,\Delta x}{\Delta t}$

but $l\,\Delta x = \Delta A$, the area 'swept out' by the conductor in a time Δt so $\varepsilon = \dfrac{B\,\Delta A}{\Delta t}$

but $B\,\Delta A = \Delta\phi$ so $\varepsilon = \dfrac{\Delta\phi}{\Delta t}$

In words, 'the e.m.f. induced is equal to the rate of cutting of flux'. If the current is kept stationary and the magnets moved, the same effect is produced.

TRANSFORMER–INDUCED E.M.F.

An e.m.f. is also produced in a wire if the magnetic field changes with time.

If the amount of flux passing through one turn of a coil is ϕ, then the total **flux linkage** with all N turns of the coil is given by

Flux linkage = $N\,\phi$

The universal rule that applies to all situations involving induced e.m.f. can now be stated as

'The magnitude of an induced e.m.f. is proportional to the rate of change of flux linkage.'

This is known as **Faraday's law**

$$\text{e.m.f.} = N\,\frac{\Delta\phi}{\Delta t}$$

INDUCED E.M.F.

The e.m.f. induced depends on:
- The speed of the wire
- The strength of the magnetic field
- The length of the wire in the magnetic field

We can calculate the magnitude of the induced e.m.f. by considering an electron at equilibrium in the middle of the wire.

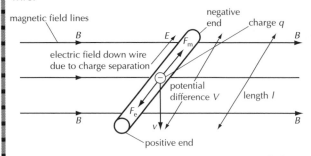

Electrical force due to e.m.f., $Fe = E \times q = \left(\dfrac{V}{l}\right) \times q$

Magnetic force due to movement, $Fm = B\,q\,v$

So $\qquad B\,q\,v = \left(\dfrac{V}{l}\right)q$

$$V = B\,l\,v$$

As no current is flowing, the e.m.f. = potential difference

$$\text{e.m.f.} = B\,l\,v$$

If the wire was part of a complete circuit (outside the magnetic field), the e.m.f. induced would cause a current to flow.

LENZ'S LAW

Lenz's law states that
'The direction of the induced e.m.f. is such that if an induced current were able to flow, it would oppose the change which caused it.'

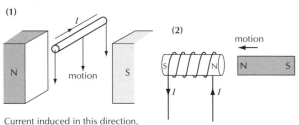

(1) Current induced in this direction, the force would be upwards (left-hand rule)
∴ original motion would be opposed.

(2) If current were induced this way, the induced field would repel the magnet – opposing motion.

EXAMPLE

An aeroplane flies at 200 m s^{-1}. Estimate the maximum p.d. that can be generated across its wings.

Vertical component of Earth's magnetic field = 10^{-5}T (approximately)

Length across wings = 30 m (estimated)

e.m.f. = $10^{-5} \times 30 \times 200$

$\qquad = 6 \times 10^{-2}$ V

$\qquad = 0.06$ V

COIL ROTATING IN A MAGNETIC FIELD – A.C. GENERATOR

The structure of a typical a.c. generator is shown below.

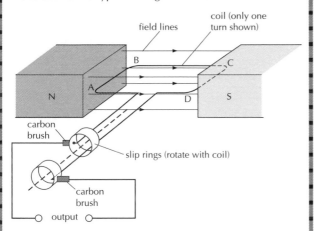

a.c. generator

The coil of wire rotates in the magnetic field due to an external force. As it rotates the flux linkage of the coil changes with time and induces an e.m.f. (Faraday's law) causing a current to flow. The sides AB and CD of the coil experience a force opposing the motion (Lenz's law). The work done rotating the coil generates electrical energy.

R.M.S. VALUES

If the output of an a.c. generator is connected to a resistor an alternating current will flow. A sinusoidal potential difference means a sinusoidal current.

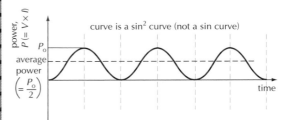

The graph shows that the average power dissipation is half the peak power dissipation for a sinusoidal current.

$$\text{Average power} = \frac{I_0^2 R}{2} = \left(\frac{I_0}{\sqrt{2}}\right)^2 R$$

Thus the effective current through the resistor is $\sqrt{}$ (mean value of I^2) and it is called the **root mean square** current or **r.m.s.** current, $I_{r.m.s.}$.

$$I_{r.m.s.} = \frac{I_0}{\sqrt{2}}$$

When a.c. values for voltage or current are quoted, it is the root mean square value that is being used. In Europe this value is 230V, whereas in the USA it is 110V.

TRANSFORMER OPERATION

An alternating potential difference is put into the transformer, and an alternating potential difference is given out. The value of the output potential difference can be changed (increased or decreased) by changing the **turns ratio**. A **step-up** transformer increases the voltage, whereas a **step-down** transformer decreases the voltage.

The following sequence of calculations provides the correct method for calculating all the relevant values.
- The output voltage is fixed by the input voltage and the turns ratio.
- The value of the load that you connect fixes the output current (using $V = I R$).
- The value of the output power is fixed by the values above ($P = V I$).
- The value of the input power is equal to the output power for an ideal transformer.
- The value of the input current can now be calculated (using $P = V I$).

So how does the transformer manage to alter the voltages in this way?

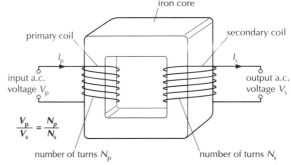

Transformer structure

- The alternating p.d. across the primary creates an a.c. within the coil and hence an alternating magnetic field in the iron core.
- This alternating magnetic field links with the secondary and induces an e.m.f. The value of the induced e.m.f. depends on the rate of change of flux linkage which increases with increased number of turns on the secondary. The input and output voltages are related by the turns ratio.

TRANSMISSION OF ELECTRICAL POWER

Transformers play a very important role in the safe and efficient transmission of electrical power over large distances.

- If large amounts of power are being distributed, then the currents used will be high. (Power = $V I$)
- The wires cannot have zero resistance. This means they must dissipate some power
- Power dissipated is $P = I^2 R$. If the current is large then the (current)^2 will be very large.

- Over large distance, the power wasted would be very significant.
- The solution is to choose to transmit the power at a very high potential difference.
- Only a small current needs to flow.
- A very high potential difference is much more efficient, but very dangerous to the user.
- Use step-up transformers to increase the voltage for the transmission stage and then use step-down transformers for the end user.

1   The **primary** of an ideal transformer has 1000 turns and the **secondary** 100 turns. The current in the primary is 2 A and the input power is 12 W.

Which **one** of the following about the **secondary current** and the **secondary power output** is true?

	secondary current	secondary power output
A	20 A	1.2 W
B	0.2 A	12 W
C	0.2 A	120 W
D	20 A	12 W

2   The diagram shows the field lines due to two small charged particles P and Q.

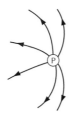 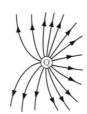

Consider the following statements:

I.   The charge on P is smaller than the charge on Q.

II. The electrostatic force on P is smaller than the force on Q.

Which of these statements is/are true?

A Only I   **C** Both I and II

B Only II   **D** Neither I nor II

3   A helium nucleus $^4_2$He and a proton $^1_1$H are both accelerated from rest through the same potential difference. The ratio of the kinetic energy of the helium nucleus to that of the proton will be

A 1:2   **B** 1:$\sqrt{2}$ **C** $\sqrt{2}$:1   **D** 2:1

4   In the arrangement shown below, if the current in the straight wire is **increasing** with time, the current induced in the loop will be

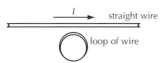

A zero.   **C** anticlockwise.

B clockwise.   **D** alternating.

5   The diagram shows a simple generator with the coil rotating between magnetic poles. Electrical contact is maintained through two brushes, each touching a slip ring.

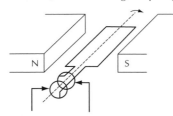

At the instant when the rotating coil is oriented as shown, the voltage across the brushes

A is zero.

B has its maximum value.

C has the same constant value as in all other orientations.

D reverses direction.

6   An alternating voltage of peak value 300 V is applied across a 50 Ω resistor. The average power dissipated in watts will be

A zero   **B** $\dfrac{(300)^2}{50}$ **C** $\dfrac{(300)^2}{50\sqrt{2}}$ **D** $\dfrac{(300)^2}{50 \times 2}$

7   Two loops of wire are next to each other as shown here. There is a source of alternating e.m.f. connected to loop 1 and an ammeter in loop 2.

The variation with time of the current in loop 1 is shown as line 1 in each of the graphs below. In which graph does line 2 best represent the current in loop 2?

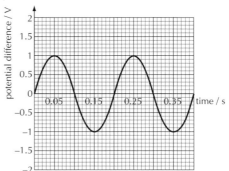

8   A loop of wire of negligible resistance is rotated in a magnetic field. A 4 Ω resistor is connected across its ends. A cathode ray oscilloscope measures the varying induced potential difference across the resistor as shown below.

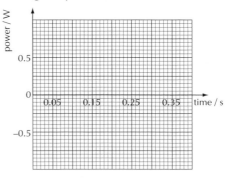

(a) If the coil is rotated at twice the speed, show on the axes above how potential difference would vary with time.   [2]

(b) What is the r.m.s. value of the induced potential difference, $v_{r.m.s.}$, at the **original** speed of rotation?   [1]

(c) On the axes below, draw a graph showing how the power dissipated in the resistor varies with time, at the **original** speed of rotation.   [3]

HL The quantum nature of radiation

PHOTOELECTRIC EFFECT

Under certain conditions, when light (ultra-violet) is shone onto a metal surface (such as zinc), electrons are emitted from the surface.

More detailed experiments (see below) showed that:

- below a certain **threshold frequency**, f_0, no photoelectrons are emitted, no matter how long one waits.
- above the threshold frequency, the maximum kinetic energy of these electrons depends on the frequency of the incident light.
- the number of electrons emitted depends on the intensity of the light and does not depend on the frequency.
- there is no noticeable delay between the arrival of the light and the emission of electrons.

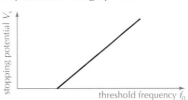

polished zinc plate

electrons emitted from surface when light is shone

radiation

negatively charged electroscope

leaf goes down when light is shone

These observations cannot be reconciled with the view that light is a wave. A wave of any frequency should eventually bring enough energy to the metal plate.

STOPPING POTENTIAL EXPERIMENT

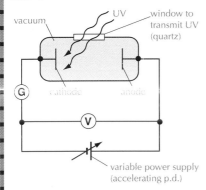

vacuum
UV
window to transmit UV (quartz)

cathode   anode

G

V

variable power supply (accelerating p.d.)

In the apparatus above, photoelectrons are emitted by the cathode. They are then accelerated across to the anode by the potential difference.

The potential between cathode and anode can also be reversed.

In this situation, the electrons are decelerated. At a certain value of potential, the stopping potential, V_s, no more photocurrent is observed. The photoelectrons have been brought to rest before arriving at the anode.

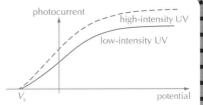

photocurrent

high-intensity UV

low-intensity UV

V_s potential

The stopping potential depends on the frequency of UV light in the linear way shown in the graph below.

stopping potential V_s

threshold frequency f_0

The stopping potential is a measure of the maximum kinetic energy of the electrons.

Max KE of electrons = $V_s e$

[since p.d. = $\dfrac{\text{energy}}{\text{charge}}$

and e = charge on an electron]

$\therefore \dfrac{1}{2} mv^2 = V_s e \quad \therefore v = \sqrt{\dfrac{2V_s e}{m}}$

EINSTEIN MODEL

Einstein introduced the idea of thinking of light as being made up of particles.

His explanation was that

- electrons at the surface need a certain minimum energy in order to escape from the surface. This minimum energy is called the **work function** of the metal and given the symbol ϕ.
- The UV light energy arrives in lots of little packets of energy – the packets are called photons.
- The energy in each packet is fixed by the frequency of UV light that is being used, whereas the number of packets arriving per second is fixed by the intensity of the source.
- The energy carried by a photon is given by

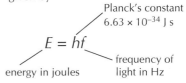

Planck's constant
6.63×10^{-34} J s

$$E = hf$$

energy in joules

frequency of light in Hz

- Different electrons absorb different photons. If the energy of the photon is large enough, it gives the electron enough energy to leave the surface of the metal.
- Any "extra" energy would be retained by the electron as kinetic energy.
- If the energy of the photon is too small, the electron will still gain this amount of energy but it will soon share it with other electrons.

Above the threshold frequency, incoming energy of photons = energy needed to leave the surface + kinetic energy

in symbols,

$hf = \phi + KE_{max}$ or $hf = \phi + V_s e$

This means that a graph of frequency against stopping potential should be a straight line of gradient $\dfrac{e}{h}$.

EXAMPLE

What is the maximum velocity of electrons emitted from a zinc surface ($\phi = 4.2$ eV) when illuminated by EM radiation of wavelength 200 nm?

$\phi = 4.2$ eV $= 4.2 \times 1.6 \times 10^{-19}$ J $= 6.72 \times 10^{-19}$ J

Energy of photon $= h\dfrac{c}{\lambda} = \dfrac{6.63 \times 10^{-34} \times 3 \times 10^8}{2 \times 10^{-7}}$

$= 9.945 \times 10^{-19}$ J

$\therefore$ KE of electron $= (9.945 - 6.72) \times 10^{-19}$ J

$= 3.225 \times 10^{-19}$ J

$\therefore v = \sqrt{\dfrac{2 \, KE}{m}}$

$= \sqrt{\dfrac{2 \times 3.225 \times 10^{-19}}{9.1 \times 10^{-31}}}$

$= 8.4 \times 10^5$ m s^{-1}

WAVE-PARTICLE DUALITY

The photoelectric effect of light waves clearly demonstrates that light can behave like particles, but its wave nature can also be demonstrated – it reflects, refracts, diffracts and interferes just like all waves. So what exactly is it? It seems reasonable to ask two questions.

1. *Is light a wave or is it a particle?*
The correct answer to this question is "yes"! At the most fundamental and even philosophical level, light is just light. Physics tries to understand and explain what it is. We do this by imagining models of its behaviour. Sometimes it helps to think of it as a wave and sometimes it helps to think

of it as a particle, but neither model is complete. Light is just light. This dual nature of light is called **wave-particle duality**.

2. *If light waves can show particle properties, can particles such as electrons show wave properties?*
Again the correct answer is "yes". Most people imagine moving electrons as little particles having a definite size, shape, position and speed. This model does not explain why electrons can be diffracted through small gaps. In order to diffract they must have a wave nature. Once again they have a dual nature. See the experiment below.

DE BROGLIE HYPOTHESIS

If matter can have wave properties and waves can have matter properties, there should be a link between the two models. The de Broglie hypothesis is that all moving particles have a "matter wave" associated with them. The wavelength of this matter wave is given by the de Broglie equation:

$$\lambda = \frac{h}{p}$$

λ – wavelength in m

h – Planck's constant – 6.63×10^{-34} Js

p – momentum in kg m s^{-1}

This matter wave can be thought of as a probability function associated with the moving particle. The (amplitude)^2 of the wave at any given point is a measure of the probability of finding the particle at that point.

ELECTRON DIFFRACTION EXPERIMENT

In order to show diffraction, an electron 'wave' must travel through a gap of the same order as its wavelength. The atomic spacing in crystal atoms provides such gaps. If a beam of electrons impinges upon powdered carbon then the electrons will be diffracted according to the wavelength.

If accelerating p.d. = 1000 V

$$\begin{aligned} \text{KE of electrons} &= eV \\ &= 1.6 \times 10^{-19} \times 1000 \text{ J} \\ &= 1.6 \times 10^{-16} \text{ J} \end{aligned}$$

For non-relativistic speeds, the link between momentum p and KE is given by

$$KE = \frac{1}{2}mv^2 = \frac{p^2}{2m}$$

$$p = \sqrt{2m \times KE}$$

$$= \sqrt{2 \times 9.1 \times 10^{-31} \times 1.6 \times 10^{-16}} \text{ kg m s}^{-1}$$

$$= 1.7 \times 10^{-23} \text{ kg m s}^{-1}$$

$$\lambda = \frac{h}{p}$$

$$= \frac{6.6 \times 10^{-34}}{1.7 \times 10^{-23}} \text{ m}$$

$$= 3.9 \times 10^{-11} \text{ m}$$

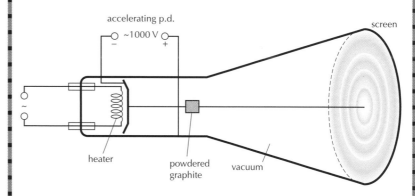

The circles correspond to the angles where constructive interference takes place. They are circles because the powdered carbon provides every possible orientation of gap. A higher accelerating potential for the electrons would result in a higher momentum for each electron. According to the de Broglie relationship, the wavelength of the electrons would thus decrease. This would mean that the size of the gaps is now proportionally bigger than the wavelength so there would be less diffraction. The circles would move in to smaller angles. The predicted angles of constructive interference are accurately verified experimentally.

Atomic spectra and atomic energy states

INTRODUCTION

As we have already seen, atomic spectra (emission and absorption) provide evidence for the quantization of the electron energy levels. Different atomic models have attempted to explain these energy levels. Two particular models of the hydrogen atom, developed by the physicists Bohr and Schrödinger, are of particular interest because their approach is completely different. In the Bohr model (historically the earlier model) the electron is considered to be a particle in orbit around the central nucleus. Schrödinger's model describes the electrons by using wave functions. Both are able to accurately predict the discrete wavelengths of the hydrogen spectrum by predicting the energy levels available.

BOHR MODEL

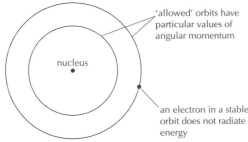

'allowed' orbits have particular values of angular momentum

nucleus

an electron in a stable orbit does not radiate energy

The details of the Bohr model are studied for option E (page 119). Essentially, Bohr took the standard "planetary" model of the hydrogen atom and filled in the mathematical details. Unlike planetary orbits, there are only a limited number of "allowed" orbits for the electron. Bohr discovered that these orbits had fixed multiples of angular momentum. The orbits were quantized in terms of angular momentum. The energy levels predicted by this quantization were in exact agreement with the discrete wavelengths of the hydrogen spectrum. Although this agreement with experiment is impressive, the model has some problems associated with it.

- The electron is assumed to be in orbit around the nucleus. This means that it is constantly accelerating (since the direction of its velocity is changing all the time). It can be demonstrated that an accelerating charge radiates energy. If this was the case for the electrons in the atom, they should be radiating energy all the time. They would spiral into the nucleus and no atoms should exist! This problem with the model is just quietly ignored – electrons in orbits are assumed not to radiate energy for some unknown reason.
- The model works to some extent for the simplest of all atoms, hydrogen. If, however, one tries to extend it to apply to other larger atoms the mathematics fail to work and the predicted energy levels fail to agree with experiment.
- The model fails to offer any explanation for the reasons behind the quantization of angular momentum. It's just an assumption that works.

SCHRÖDINGER'S MODEL

If electrons can be described by waves, then the discrete energy levels that exist should be related to different possible wave descriptions. In Schrödinger's model (page 120) the electron wave functions are essentially standing waves that fit the boundary conditions in the atom. The situation is very hard to imagine as we are talking about a wave in all three dimensions. It is probably better to simplify the situation by considering only one dimension and start by remembering the physical standing waves that can be fitted onto a stretched string (see Topic 4).

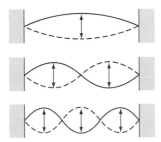

Standing waves possible on a string

The standing waves on a string have a fixed wavelength but for energy reasons the same is not true for the electron wave functions. As an electron moves away from the nucleus it must lose kinetic energy because they have opposite charges. Lower kinetic energy means that it would be travelling with a lower momentum and the de Broglie relationship predicts a longer wavelength. This means that the possible wave functions that fit the boundary conditions have particular shapes

The wave function provides a way of working out the probability of finding an electron at that particular radius. The (amplitude)^2 of the wave at any given point is a measure of the probability of finding the electron at that distance away from the nucleus – in any direction.

The wave function exists in all three dimensions, which makes it hard to visualise. Often the electron orbital is pictured as a cloud. The exact position of the electron is not known but we know where it is more likely to be.

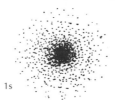

1s

Electron cloud for the 1s orbital in hydrogen

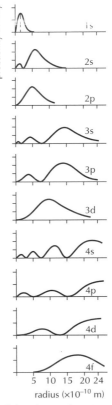

probability density

1s

2s

2p

3s

3p

3d

4s

4p

4d

4f

5 10 15 20 24
radius ($\times 10^{-10}$ m)

Probability density functions for some orbitals in the hydrogen atom. The scale on the vertical axis is different from graph to graph.

HL X-rays

PRODUCTION OF X-RAYS

At one time or another in their lives, most people have an x-ray photograph taken of some part of their body. The experimental procedure used for the production of x-rays relies on electrons.

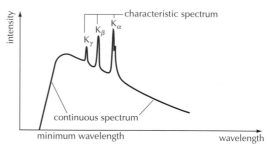

EHT (extra-high tension)

V
− (accelerating p.d.) +

concave focusing cathode

cooled copper block anode

tungsten target

heater ends

emitting filament

X-rays

adjustable diaphragm

vacuum

lead shield

X-ray tube

- Electrons are accelerated by travelling through a potential difference, V. The kinetic energy gained is thus $V\,e$.
- The fast moving electrons collide with the metal target and the collision results in x-rays. There is also a great deal of heat generated in the target. This needs to be kept cool by being kept rotating in oil.
- The adjustable diaphragm is used to define the beam of x-rays.

EXAMPLE

Calculate the minimum wavelength x-rays produced by an electron beam accelerated through a potential difference of 30 keV.

$$\text{Energy} = Ve$$
$$= 30\,000 \times 1.6 \times 10^{-19}$$
$$= 4.8 \times 10^{-15}\,\text{J}$$
$$\therefore\ hf = 4.8 \times 10^{-15}$$
$$f = \frac{4.8 \times 10^{-15}}{6.63 \times 10^{-34}}\ \text{Hz}$$
$$= 7.24 \times 10^{18}\ \text{Hz}$$
$$\lambda = \frac{c}{f}$$
$$= \frac{3 \times 10^{8}}{7.24 \times 10^{18}}$$
$$= 4.1 \times 10^{-11}\ \text{m}$$

TYPICAL X-RAY SPECTRUM

The x-rays emitted from the target contain a range of wavelengths – all with different relative amplitudes. A graph of a typical x-ray spectrum is

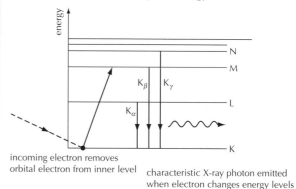

characteristic spectrum

K_γ K_β K_α

intensity

continuous spectrum

minimum wavelength

wavelength

There are two different process taking place in order for this spectrum to be produced

- **Continuous features**
  As the incoming electrons collide with the target atoms, they are decelerated. This deceleration means that X-rays are emitted. The energy of the x-ray photon depends on the energy lost in the collisions. The maximum amount of energy that can be lost is all the initial kinetic energy of the electrons. The maximum energy available means that there is a maximum frequency of x-rays produced. This corresponds to a minimum wavelength limit shown on the graph.

- **Characteristic features**
  In some circumstances, the collisions between the incoming electrons and the target atoms can cause electrons from the inner orbital of the target atom to be promoted up to higher energy levels. When these electrons fall back down they emit x-rays of a particular frequency which is fixed by the energy levels available.

energy

N

M

K_β K_γ

L

K_α

K

incoming electron removes orbital electron from inner level

characteristic X-ray photon emitted when electron changes energy levels

These two processes combine to give the overall spectrum. Changes to this spectrum can thus be predicted.

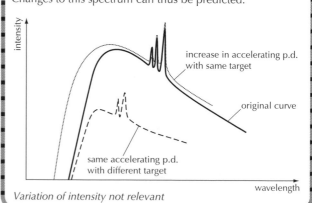

intensity

increase in accelerating p.d. with same target

original curve

same accelerating p.d. with different target

wavelength

Variation of intensity not relevant

ⒽⓁ The nucleus

THE NUCLEUS – SIZE

In the example below, alpha particles are allowed to bombard gold atoms.

As they approach the gold nucleus, they feel a force of repulsion. If an alpha particle is heading directly for the nucleus, it will be reflected straight back along the same path. It will have got as close as it can. Note that none of the alpha particles actually collides with the nucleus – they do not have enough energy.

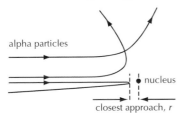

Alpha particles are emitted from their source with a known energy. As they come in they gain electrostatic potential energy and lose kinetic energy (they slow down). At the closest approach, the alpha particle is temporarily stationary and all its energy is potential.

Since electrostatic energy = $\frac{q_1 q_2}{4\pi\varepsilon_0 r}$, and we know q_1, the charge on an alpha particle and q_2, the charge on the gold nucleus we can calculate r.

ENERGY LEVELS

The energy levels in a nucleus are higher than the energy levels of the electrons but the principle is the same. When an alpha particle or a gamma photon is emitted from the nucleus only discrete energies are observed. These energies correspond to the difference between two **nuclear energy levels** in the same way that the photon energies correspond to the difference between two **atomic energy levels**.

Beta particles are observed to have a continuous spectrum of energies. In this case there is another particle (the antineutrino in the case of beta minus decay) that shares the energy. Once again the amount of energy released in the decay is fixed by the difference between the *nuclear* energy levels involved. The beta particle and the antineutrino can take varying proportions of the energy available. The antineutrino, however, is very difficult to observe (see next section).

The decay of ^{226}Ra into ^{222}Rn

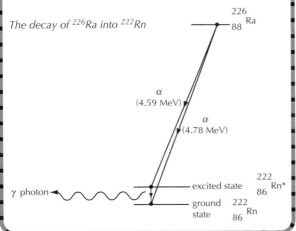

EXAMPLE

If the α particles have an energy of 4.2 MeV, the closest approach to the gold nucleus ($Z = 79$) is given by

$$\frac{(2 \times 1.6 \times 10^{-19})(79 \times 1.6 \times 10^{-19})}{4 \times \pi \times 8.85 \times 10^{-12} \times r} = 4.2 \times 10^6 \times 1.6 \times 10^{-19}$$

$$\therefore r = \frac{2 \times 1.6 \times 10^{-19} \times 79}{4 \times \pi \times 8.85 \times 10^{-12} \times 4.2 \times 10^6}$$

$$= 5.4 \times 10^{-14} \text{ m}$$

MASS

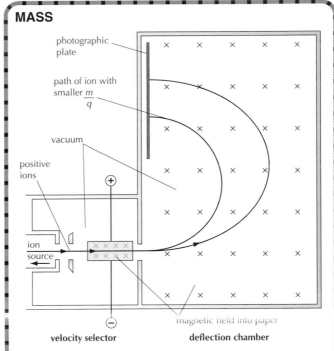

Essentials of a mass spectrometer

Isotopes of the same element have the same chemical properties so they cannot be separated using chemical reactions. The mass spectrometer provides a way of determining the masses of individual nuclei. The principle is to use a magnetic field to deflect moving ions of a substance. If a moving ion enters a constant magnetic field, B, it will follow a circular path (see Topic 5). The magnetic force provides the centripetal force required.

$$Bqv = \frac{mv^2}{r}$$

The radius of the circle is thus $r = \frac{mv}{Bq}$

If the ions have the same charge, q, and they are all selected to be travelling at the same speed, v, then the radius of the circle will be depend on the mass of the ion. A larger-mass ion will travel in a larger circle. The mass spectrometer can thus easily demonstrate that there are different values of nuclear mass for a given element. In other words, isotopes exist.

Ⓗ Radioactive decay

NEUTRINOS AND ANTINEUTRINOS

Understanding beta decay properly requires accepting the existence of a virtually undetectable particle, the neutrino. It is needed to account for the "missing" energy and (angular) momentum when analysing the decay mathematically. Calculations involving mass difference mean that we know how much energy is available in beta decay. For example, an isotope of hydrogen, tritium, decays as follows:

$$^3_1H \rightarrow ^3_2He + ^0_{-1}\beta$$

The mass difference for the decay is 19.5 k eV c^{-2}. This means that the beta particles should have 19.5 k eV of kinetic energy. In fact, a few beta particles are emitted with this energy, but all the others have less than this. The average energy is about half this value and there is no accompanying

gamma photon. All beta decays seemed to follow a similar pattern.

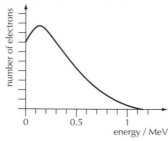

The energy distribution of the electrons emitted in the beta decay of bismuth-210. The kinetic energy of these electrons is between zero and 1.17 MeV.

The neutrino (and antineutrino) must be electrically neutral. Its mass would have to be very small, or even zero. It carries away the excess energy but it is very hard to detect. One of the triumphs of the particle physics of the last century was to be able to design

experiments that confirmed its existence. The full equation for the decay of tritium is:

$$^3_1H \rightarrow ^3_2He + ^0_{-1}\beta + \bar{v}$$

where $\bar{v}$ is an antineutrino

As has been mentioned before, another form of radioactive decay can also take place, namely positron decay. In this decay, a proton within the nucleus decays into a neutron and the antimatter version of an electron, a positron, is emitted.

$$^1_1p \rightarrow ^1_0n + ^0_{+1}\beta^+ + v$$

In this case, the positron, β^+, is accompanied by a neutrino.

The antineutrino is the antimatter form of the neutrino.

e.g. $^{19}_{10}Ne \rightarrow ^{19}_9F + ^0_{+1}\beta^+ + \bar{v}$

$^{14}_6C \rightarrow ^{14}_7N + ^0_{-1}\beta^- + \bar{v}$

MATHEMATICS OF EXPONENTIAL DECAY

The basic relationship that defines exponential decay as a random process is expressed as follows:

$$\frac{dN}{dt} \propto -N$$

The constant of proportionality between the rate of decay and the number of nuclei available to decay is called the decay constant and given the symbol λ. Its units are time^{-1} i.e. s^{-1} or yr^{-1} etc.

$$\frac{dN}{dt} = -\lambda N$$

The solution of this equation is:

$$N = N_0e^{-\lambda t}$$

The activity of a source, A, is:

$$A = A_0e^{-\lambda t}$$

It is useful to take natural logarithms:

$$\ln(N) = \ln(N_0e^{-dt})$$
$$= \ln(N_0) + \ln(e^{-dt})$$
$$= \ln(N_0) - dt\ln(e)$$
$$\therefore \ln(N) = \ln(N_0) - dt \quad (\text{since } \ln(e) = 1)$$

This is of the form $y = c + mx$ so a graph of $\ln N$ vs. t will give a straight-line graph.

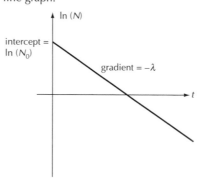

If
$$N = N_0e^{-\lambda t}$$
$$t = \frac{T_{\frac{1}{2}}}{}$$
$$N = \frac{N_0}{2}$$
so
$$\frac{N_0}{2} = N_0e^{-\lambda T_{\frac{1}{2}}}$$
$$\therefore \frac{1}{2} = e^{-\lambda T_{\frac{1}{2}}}$$
$$\therefore \ln\left(\frac{1}{2}\right) = -\lambda T_{\frac{1}{2}}$$
$$\therefore \lambda T_{\frac{1}{2}} = -\ln\left(\frac{1}{2}\right)$$
$$= \ln 2$$
$$\therefore T_{\frac{1}{2}} = \frac{\ln 2}{\lambda}$$

METHODS FOR MEASURING HALF-LIFE

When measuring the activity of a source, the background rate should be subtracted.

- If the half-life is short, then readings can be taken of activity against time.
  → A simple graph of activity against time would produce the normal exponential shape. Several values of half-life could be read from the graph and then averaged. This method is simple and quick but not the most accurate.
  → A graph of ln (activity) against time could be produced. This should give a straight line and the decay constant can be calculated from the gradient.
- If the half-life is long, then the activity will effectively be constant over a period of time. In this case one needs to find a way to calculate the number of nuclei present and then use

$$\frac{dN}{dt} = -\lambda N.$$

EXAMPLE

The half-life of a radioactive isotope is 10 days. Calculate the fraction of a sample that remains after 25 days.

$T_{\frac{1}{2}} = 10$ days

$$\lambda = \frac{\ln 2}{T_{\frac{1}{2}}}$$
$$= 6.93 \times 10^{-2} \text{ day}^{-1}$$
$$N = N_0e^{-\lambda t}$$

Fraction remaining $= \frac{N}{N_0}$

$$= e^{-(6.92 \times 10^{-2} \times 25)}$$
$$= 0.187$$
$$= 18.7\%$$

Particle Physics

ANTIMATTER – PARTICLE PRODUCTION AND ANNIHILATION

In a particle accelerator particles are accelerated up to very high energies and then collided. New particles can be created in accordance with Einstein's mass-energy relation. The tracks of the particles are recorded and the curvature of the path (as a result of a magnetic field) is used to calculate the mass and the velocity and the charge of the particle.

All particles have their antimatter equivalent which has the same mass but opposite properties such as charge etc. When matter and antimatter combine the result is annihilation of the mass and its conversion into energy – a photon. The reverse process is also observed, mass can be created from energy.

CLASSIFICATION OF PARTICLES

Particle accelerator experiments identify many, many "new" particles. Two original classes of particles were identified – the **leptons** (= "light") and the **hadrons** (= "heavy"). The hadrons were subdivided into **mesons** and **baryons**. Another class of particles is involved in the mediation of the interactions between the particles. These were called **bosons** or "exchange bosons".

CONSERVATION LAWS

Not all reactions between particles are possible. The study of the reactions that did take place gave rise to some experimental conservation laws that applied to particle physics. Some of these laws were simply confirmation of conservation laws that were already known to physicists – charge, momentum (linear and angular) and mass-energy. On top of these fundamental laws there appeared to be other rules that were never broken e.g. the law of conservation of baryon number. If all baryons were assigned a "baryon number" of 1 (and all antibaryons were assigned a baryon number of -1) then the total number of baryons before and after a collision was always the same. A similar law of conservation of lepton number applies.

Other reactions suggested new and different particle properties that were often, but not always, conserved in reactions. "Strangeness" and "charm" are examples of two such properties. Strangeness is conserved in all strong interactions.

CLASSIFICATION OF INTERACTIONS- EXCHANGE PARTICLES

There are only four fundamental interactions that exist: Gravity, Electromagnetic, Strong and Weak.

- At the end of the nineteenth century, Maxwell showed that the electrostatic force and the magnetic force were just two different aspects of the more fundamental electromagnetic force.
- Friction is simply a result of the forces between atoms and this must be governed by electromagnetic interactions.
- The strong and the weak interaction only exist over nuclear ranges.
- The strong force binds the nucleus together.
- The weak force explains radioactive β decay.

- The electromagnetic force and the weak nuclear force are now considered to be aspects of a single electroweak force.
- All four interactions can be thought of as being mediated by an exchange of particles. Each interaction has its own exchange particle or particles. The bigger the mass of the exchange boson, the smaller the range of the force concerned.

Interaction	Relative strength	Range (m)	Exchange particle
strong	1	~10^{-15}	8 different gluons
electromagnetic	10^{-2}	infinite	photon
weak	10^{-13}	~10^{-17}	w^+, w^-, z^0
gravity	10^{-39}	infinite	graviton

QUARK MODEL

All hadrons are made up from different combinations of fundamental particles called quarks. There are six different types of quark and six types of antiquark. This very neatly matches the six leptons that are also known to exist. Quarks are affected by the strong force (see below), whereas leptons are not. The weak interaction can change one type of quark into another.

	Electric charge	'Generation' 1	2	3
Quarks	$+\frac{2}{3}$	u (up) $M = 5$ MeV	c (charm) $M = 1500$ MeV	t (top) $M = 174\,000$ MeV
	$-\frac{1}{3}$	d (down) $M = 10$ MeV	s (strange) $M = 200$ MeV	b (bottom) $M = 4700$ MeV
Leptons	0	ν_e (electron-neutrino) $M = 0$ or almost 0	ν_μ (muon-neutrino) $M = 0$ or almost 0	ν_τ (tau-neutrino) $M = 0$ or almost 0
	-1	e (electron) $M = 0.511$ MeV	μ (muon) $M = 105$	τ (tau) $M = 1784$ MeV

Isolated quarks cannot exist. They can exist only in twos or threes. Mesons are made from two quarks (a quark and an antiquark) whereas baryons are made up of a combination of three quarks.

	Name of particle	Quark structure
Baryons	proton (p)	u u d
	neutron (n)	u d d
Mesons	π^-	d ū
	π^+	u d̄

The force between quarks is still the strong interaction but the full description of this interaction is termed QCD theory – quantum chromodynamics. A quark can be one of three different "colours" – red, green or blue. Only the "white" combinations are possible. The force between quarks is sometimes called the colour force. Eight different types of gluon mediate it.

1 The diameter of a proton is of the order

    A 10^{-9} m   **B** 10^{-11} m   **C** 10^{-13} m   **D** 10^{-15} m

2 When bullets leave the barrel of a rifle they are not observed to diffract because

    A the de Broglie hypothesis only applies to electrons.

    B the speed of the bullet is not great enough.

    C the de Broglie wavelength of the bullet is too small.

    D the de Broglie wavelength of the bullet is too large.

3 When a particle and its anti-particle undergo a mutual annihilation, which one of the following need **not** be conserved?

    A Kinetic energy   **C** Electric charge

    B Linear momentum   **D** Angular momentum

4 The diagram represents the available energy levels of an atom. How many emission lines could result from electron transitions between these energy levels?

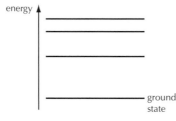

    A 3   **B** 6   **C** 8   **D** 12

5 The accelerating voltage in an X-ray tube is increased. Consider the following three statements about this situation.

    I   The intensity of the X radiation produced is increased.

    II   The X radiation becomes more penetrating.

    III The short wavelength limit is reduced.

    Which of the above statements are correct?

    A I and II   **C** I and III

    B II and III   **D** I, II and III

6 A medical physicist wishes to investigate the decay of a radioactive isotope and determine its decay constant and half-life. A Geiger-Müller counter is used to detect radiation from a sample of the isotope, as shown.

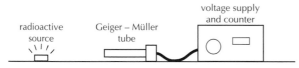

    (a) Define the activity of a radioactive sample.   [1]

    Theory predicts that the activity A of the isotope in the sample should decrease exponentially with time t according to the equation $A = A_0 e^{-\lambda t}$, where A_0 is the activity at $t = 0$ and λ is the decay constant for the isotope.

    (b) Manipulate this equation into a form which will give a straight line if a semi-log graph is plotted with appropriate variables on the axes. State what variables should be plotted.   [2]

    The Geiger-counter detects a proportion of the particles emitted by the source. The physicist records the count-rate R of particles detected as a function of time t and plots the data as a graph of ln R versus t, as shown below.

    (c) Does the plot show that the experimental data are consistent with an *exponential* law? Explain.   [1]

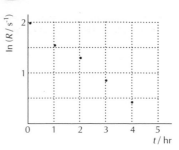

    (d) The Geiger-counter does not measure the total activity A of the sample, but rather the count-rate R of those particles that enter the Geiger tube. Explain why this will not matter in determining the decay constant of the sample.   [1]

    (e) From the graph, determine a value for the decay constant λ.   [2]

    The physicist now wishes to calculate the half-life.

    (f) Define the half-life of a radioactive substance.   [1]

    (g) Derive a relationship between the decay constant λ and the half-life τ.   [2]

    (h) Hence calculate the half-life of this radioactive isotope.   [1]

7 Two protons within a nucleus can feel an electrostatic repulsion while at the same time feeling an attraction due to the strong force. If their separation is increased to 10^{-14}m, the magnitude of both forces change.

    (a) Explain how each force (strong and electrostatic) can be viewed in terms of the exchange of virtual particles.   [3]

    (b) Explain how each force varies as the separation is increased.   [2]

8 The following two tables summarise some of the properties of the neutron, proton, electron and antineutrino and of two quarks. (You may find some of the data useful in answering this question.)

particle name	charge	baryon no.	lepton no.
neutron (n)	0	1	0
proton (p$^+$)	1	1	0
electron (e$^-$)	−1	0	1
antineutrino (v̄)	0	0	−1

quark name	charge	baryon no.	lepton no.
up (u)	2/3	1/3	0
down (d)	−1/3	1/3	0

    (a) What are the quarks making up a neutron?   [2]

    (b) This diagram shows the decay of a free neutron as mediated by the intermediate, short-lived W particle, which is represented by the wavy line.

        (i)   What fundamental interaction is involved in this decay process?   [1]

        (ii) What must be the charge, baryon number and lepton number of the W particle?   [3]

        (iii) In terms of the current quark model of neucleon structure, what changes have taken place in the internal make up of the neutron to change it into a proton?   [2]

Energy sources and power generation

RENEWABLE / NON-RENEWABLE ENERGY SOURCES

The law of conservation of energy states that energy is neither created or destroyed, it just changes form. As far as human societies are concerned, if we wish to use devices that require the input of energy, we need to identify sources of energy. **Renewable** sources of energy are those that cannot be used up, whereas **non-renewable** sources of energy can be used up and eventually run out.

Renewable sources	Non-renewable sources
hydroelectrical	coal
photovoltaic cells	oil
active solar heaters	natural gas
wind	nuclear
biofuels	

Sometimes the sources are hard to classify so care needs to be taken when deciding whether a source is renewable or not. One point that sometimes worries students is that the Sun will eventually run out as a source of energy for the Earth, so no source is perfectly renewable! This is true, but all of these sources are considered from the point of view of life on Earth. When the Sun runs out, then so will life on Earth. Other things to keep in mind include:
• Nuclear sources (both fission and fusion) consume a material as their source so they must be non-renewable. On the other hand, the supply available can make the source **effectively** renewable.

• It is possible for a fuel to be managed in a renewable or a non-renewable way. For example, if trees are cut down as a source of wood to burn then this is clearly non-renewable. It is, however, possible to replant trees at the same rate as they are cut down. If this is properly managed, it could be a renewable source of energy.

Of course these possible sources must have got their energy from somewhere in the first place. Most of the energy used by humans can be traced back to energy radiated from the Sun, but not quite all of it. Possible sources are:
• the Sun's radiated energy
• gravitational energy of the Sun and the Moon
• nuclear energy stored within atoms
• the Earth's internal heat energy

Although you might think that there are other sources of energy, the above list is complete. Many everyday sources of energy (such as coal or oil) can be shown to have derived their energy from the Sun's radiated energy. On the industrial scale, electrical energy needs to be generated from another source. When you plug anything electrical into the mains electricity you have to pay the electricity-generating company for the energy you use. In order to provide you with this energy, the company must be using one (or more) of the original list of sources.

ELECTRICAL POWER PRODUCTION

In all electrical power stations the process is essentially the same. A fuel is used to release thermal energy. This thermal energy is used to boil water to make steam. The steam is used to turn turbines and the motion of the turbines is used to generate electrical energy. Transformers alter the potential difference (see page 79).

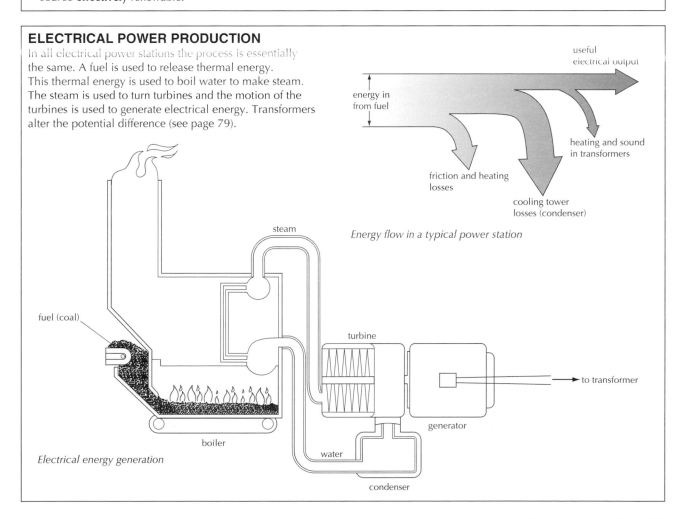

Energy flow in a typical power station

Electrical energy generation

Nuclear power stations

PRINCIPLES OF ENERGY PRODUCTION

Many nuclear power stations use uranium as the 'fuel'. This fuel is not burned – the release of energy is achieved using a fission reaction. An overview of this process is described on page 54. In each individual reaction, an incoming neutron causes a uranium nucleus to split apart. The fragments are moving fast. In other words the temperature is very high. Amongst the fragments are more neutrons. If these neutrons go on to initiate further reactions then a chain reaction is created.

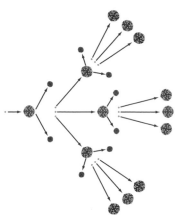

The design of a nuclear reactor needs to ensure that, on average, only one neutron from each reaction goes on to initiate a further reaction. If more reactions took place then the number of reactions would increase all that time and the chain reaction would run out of control. If fewer reactions took place, then the number of reactions would be decreasing and the fission process would soon stop.

The chance that a given neutron goes on to cause a fission reaction depends on several factors. Two important ones are:
• the number of potential nuclei 'in the way'.
• the speed (or the energy) of the neutrons.

As a general trend, as the size of a block of fuel increases so do the chances of a neutron causing a further reaction (before it is lost from the surface of the block). As the fuel is assembled together a stage is reached when a chain reaction can occur. This happens when a so-called **critical mass** of fuel has been assembled. The exact value of the critical mass depends on the exact nature of the fuel being used and the shape of the assembly.

There are particular neutrons energies that make them more likely to cause nuclear fission. In general, the neutrons created by the fission process are moving too fast to make reactions likely. Before they can cause further reactions the neutrons have to be slowed down.

MODERATOR, CONTROL RODS AND HEAT EXCHANGER

Three important components in the design of all nuclear reactors are the **moderator**, the **control rods** and the **heat exchanger**.
• Collisions between the neutrons and the nuclei of the moderator slow them down and allow further reactions to take place.
• The control rods are movable rods that readily absorb neutrons. They can be introduced or removed from the reaction chamber in order to control the chain reaction.
• The heat exchanger allows the nuclear reactions to occur in a place that is sealed off from the rest of the environment. The reactions increase the temperature in the core. This thermal energy is transferred to water and the steam that is produced turns the turbines.

A general design for one type of nuclear reactor (PWR or Pressurized Water Reactor) is shown here. It uses water as the moderator and as a coolant.

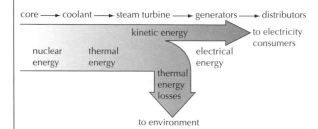

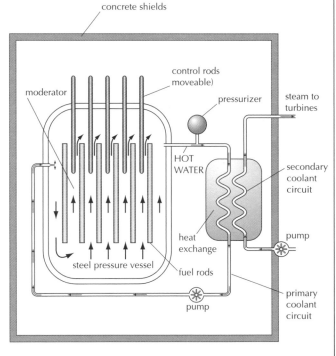

Pressurized-water nuclear reactor (PWR)

ADVANTAGES AND DISADVANTAGES

Advantages
• Extremely high 'energy density' – a great deal of energy is released from a very small mass of uranium
• reserves of uranium large compared to oil

Disadvantages
• Process produces radioactive nuclear waste that is currently just stored
• larger possible risk if anything should go wrong
• Nonrenewable (but should last a long time)

Fossil fuel power

ORIGIN OF FOSSIL FUELS

Coal, oil and natural gas are known as **fossil fuels**. These fuels have been produced over a time scale that involves tens or hundreds of millions of years from accumulations of dead matter. This matter has been converted into fossil fuels by exposure to the very high temperatures and pressure that exist beneath the Earth's surface.

Coal is formed from the dead plant matter that used to grow in swamps. Layer upon layer of decaying matter decomposed. As it was buried by more plant matter and other substances, the material became more compressed. Over the geological time-scale this turned into coal.

Oil is formed in a similar manner from the remains of microscopic marine life. The compression took place under the sea. Natural gas, as well as occurring in underground pockets, can be obtained as a by-product during the production of oil. It is also possible to manufacture gas from coal.

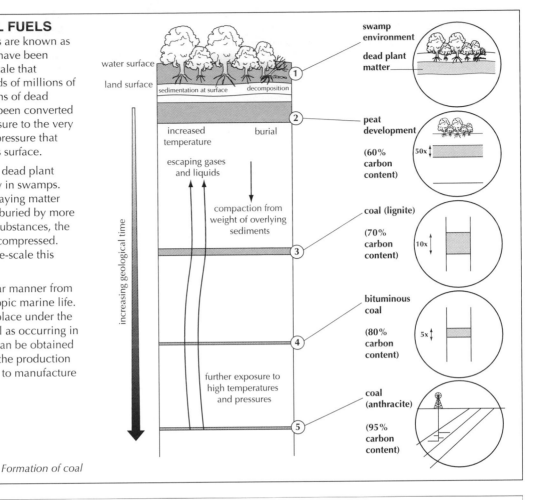

Formation of coal

ENERGY TRANSFORMATIONS

Fossil fuel power stations release energy in fuel by burning it. The thermal energy is then used to convert water into stream that once again can be used to turn turbines. Since all fossil fuels were originally living matter, the original source of this energy was the Sun. For example, millions of years ago energy radiated from the Sun was converted (by photosynthesis) into living plant matter. Some of this matter has eventually been converted into coal.

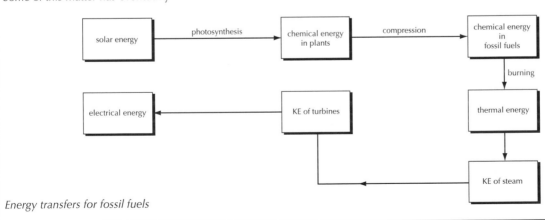

Energy transfers for fossil fuels

ADVANTAGES AND DISADVANTAGES

Advantages
- Very high 'energy density' – a great deal of energy is released from a small mass of fossil fuel
- Fossil fuels are relatively easy to transport
- Still cheap when compared to other sources of energy
- Can be used directly in the home to provide heating

Disadvantages
- Combustion products can produce pollution, notably acid rain
- Combustion products contain 'greenhouse' gases
- Extraction of fossil fuels can damage the environment
- Nonrenewable

Hydroelectric power

ENERGY TRANSFORMATIONS

The source of energy in a hydroelectric power station is the gravitational potential energy of water. If water is allowed to move downhill, the flowing water can be used to generate electrical energy.

The water can gain its gravitational potential energy in several ways.

- As part of the 'water cycle', water can fall as rain. It can be stored in large reservoirs as high up as is feasible.

- Tidal power schemes trap water at high tides and release it during a low tide.
- Water can be pumped from a low reservoir to a high reservoir. Although the energy used to do this pumping must be more than the energy regained when the water flows back down hill, this 'pumped storage' system provides one of the few large-scale methods of storing energy.

gravitational PE of water → KE of water → KE of turbines → electrical energy

ADVANTAGES AND DISADVANTAGES

Advantages
- Very 'clean' production – no harmful chemical by-products
- Renewable source of energy
- Source of energy is free

Disadvantages
- Can only be utilized in particular areas
- Construction of dam will involve land being buried under water

Wind power

ENERGY TRANSFORMATIONS

There is a great deal of kinetic energy involved in the winds that blow around the Earth. The original source of this energy is, of course, the Sun. Different parts of the atmosphere are heated to different temperatures. The temperature differences cause pressure differences, due to hot air rising or cold air sinking, and thus air flows as a result.

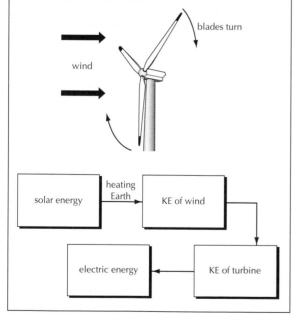

MATHEMATICS

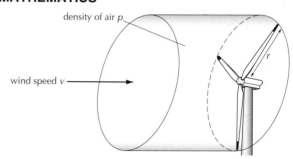

The area 'swept out' by the blades of the turbine $= A = \pi r^2$

In one second the volume of air that passes the turbine $= v A$

So mass of air that passes the turbine one second $= v A \rho$

Kinetic energy m available per second $= \frac{1}{2} m v^2$

$$= \frac{1}{2} (v A \rho) v^2$$

$$= \frac{1}{2} A \rho v^3$$

In other words, power available $= \frac{1}{2} A \rho v^3$

In practice, the kinetic energy of the incoming wind is easy to calculate, but it cannot all be harnessed – in other words the wind turbine cannot be one hundred per cent efficient. A doubling of the wind speed would mean that the available power would increase by a factor of eight.

ADVANTAGES AND DISADVANTAGES

Advantages
- Very 'clean' production – no harmful chemical by-products
- Renewable source of energy
- Source of energy is free

Disadvantages
- Source of energy is unreliable – could be a day without wind
- Low energy density – a very large area would be needed to be covered for a significant amount of energy
- Some consider large wind generators to spoil the countryside
- Can be noisy
- Best positions for wind generators are often far from centres of population

Solar power

ENERGY TRANSFORMATIONS (TWO TYPES)

There are two ways of harnessing the radiated energy that arrives at the Earth's surface from the Sun.

A **photovoltaic cell** (otherwise known as a solar cell or photocell) converts a portion of the radiated energy directly into a potential difference ('voltage'). It uses a piece of semiconductor to do this. Unfortunately, a typical photovoltaic cell produces a very small voltage and it is not able to provide much current. They are used to run electrical devices that do not require a great deal of energy. Using them in series would generate higher voltages and several in parallel can provide a higher current.

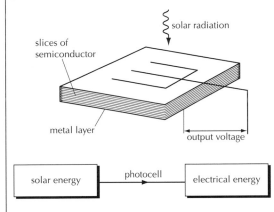

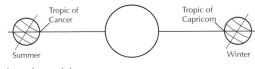

An **active solar heater** (otherwise known as a solar panel) is designed to capture as much thermal energy as possible. The hot water that it typically produces can be used domestically and would save on the use of electrical energy.

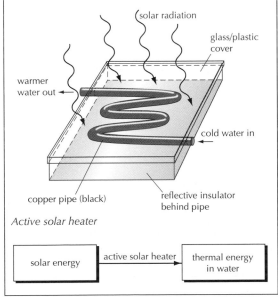

Active solar heater

SOLAR CONSTANT

The amount of power that arrives from the Sun is measured by the solar constant. It is properly defined as the amount of solar energy that falls per second on an area of 1 m² above the Earth's atmosphere that is at right angles to the Sun's rays. Its average value is about 1400 W m⁻².

This is not the same as the power that arrives on 1 m² of the Earth's surface. Scattering and absorption in the atmosphere means that often less than half of this arrives at the Earth's surface. The amount that arrives depends greatly on the weather conditions.

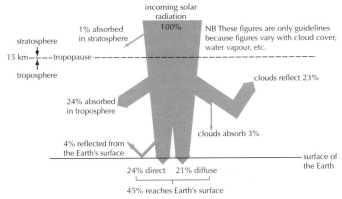

Fate of incoming radiation

Different parts of the Earth's surface (regions at different latitudes) will receive different amounts of solar radiation. The amount received will also vary with the seasons since this will affect how spread out the rays have become.

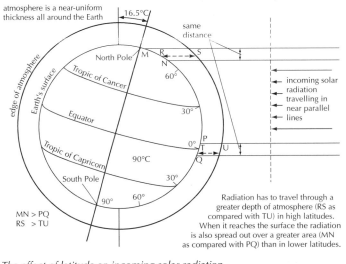

The effect of latitude on incoming solar radiation

The Earth's orbit and the seasons

ADVANTAGES AND DISADVANTAGES

Advantages
- Very 'clean' production – no harmful chemical by-products
- Renewable source of energy
- Source of energy is free

Disadvantages
- Can only be utilized during the day
- Source of energy is unreliable – could be a cloudy day
- Low energy density – a very large area would be needed for a significant amount of energy

IB QUESTIONS – OPTION C – ENERGY EXTENSION

1 A wind generator converts wind energy into electric energy. The source of this wind energy can be traced back to solar energy arriving at the Earth's surface.

  (a) Outline the energy transformations involved as solar energy converts into wind energy. [2]

  (b) List **one** advantage and **one** disadvantage of the use of wind generators. [2]

  The expression for the maximum theoretical power, P, available from a wind generator is

  $$P = \frac{1}{2} A \rho v^3$$

  where A is the area swept out by the blades,
  ρ is the density of air and
  v is the wind speed.

  (c) Calculate the maximum theoretical power, P, for a wind generator whose blades are 30 m long when a 20 m s^{-1} wind blows. The density of air is 1.3 kg m^{-3}. [2]

  (d) In practice, under these conditions, the generator only provides 3 MW of electrical power.
   (i) Calculate the efficiency of this generator. [2]
   (ii) Give **two** reasons explaining why the actual power output is less than the maximum theoretical power output. [2]

2 This question is about energy sources.

  (a) Give **one** example of a renewable energy source and **one** example of a non-renewable energy source and explain why they are classified as such. [4]

  (b) A wind farm produces 35 000 MWh of energy in a year. If there are ten wind turbines on the farm show that the average power output of **one** turbine is about 400 kW. [3]

  (c) State **two** disadvantages of using wind power to generate electrical power. [2]

3 This question is about energy transformations.

  It is said that all our energy, in whatever form, comes from the Sun.

  Wind power can be used to generate electrical energy.

  Construct an energy flow diagram which shows the energy transformations, starting with solar energy and ending with electrical energy, generated by windmills. Your diagram should indicate where energy is degraded. [7]

  > SOLAR
  > ENERGY

  > ELECTRICAL
  > ENERGY

4 This question is about a coal-fired power station which is water cooled.

  Data:

Electrical power output from the station	= 200 MW
Temperature at which water enters cooling tower	= 288 K
Temperature at which water leaves cooling tower	= 348 K
Rate of water flow through tower	= 4000 kg s^{-1}
Energy content of coal	= 2.8 × 10^7 J kg^{-1}
Specific heat of water	= 4200 J kg^{-1} K^{-1}

  Calculate

  (a) the energy per second carried away by the water in the cooling tower; [2]

  (b) the energy per second produced by burning the coal; [2]

  (c) the overall efficiency of the power station; [2]

  (d) the mass of coal burnt each second. [1]

5 This question is about tidal power systems.

  (a) Describe the principle of operation of such a system. [2]

  (b) Outline **one** advantage and **one** disadvantage of using such a system. [2]

  (c) A small tidal power system is proposed. Use the data in the table below to calculate the total energy available and hence estimate the useful output power of this system.

Height between high tide and low tide	4 m
Trapped water would cover an area of	1.0 × 10^6 m^2
Density of water	1.0 × 10^3 kg m^{-3}
Number of tides per day	2

  [4]

6 This question is about a hydroelectric power scheme using tidal energy.

  The diagram shows a hydroelectric scheme constructed in the ocean near the shore. Built into the dam wall is a system of pipes and adjustable valves (not shown), to allow water to flow one way or the other, and a turbine connected to an electric generator.

  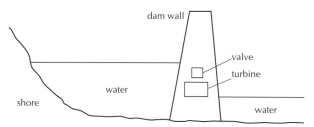

  (a) Explain in some detail how such a system might work to provide electrical energy from the tides. [5]

  (b) Tidal systems work only on a small scale and in certain places. Suggest **two** factors which make it impractical for such systems to provide electrical energy on a large scale or a widespread scale. [2]

  (c) The tides are the immediate source of energy for this hydroelectric system.
   (i) Where does the energy of the tides come from? [1]
   (ii) Discuss briefly whether tidal energy systems give us something for nothing and whether the source of tidal energy can eventually be used up. [2]

Scaling

PRINCIPLES OF SCALING

What does it mean to double the size of a block?

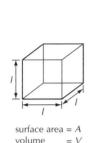
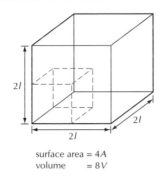

surface area = A
volume   = V

surface area = $4A$
volume   = $8V$

If all the dimensions double, then the surface area must increase by a factor of 4 and the volume by a factor of 8. Another way of saying this is that the surface area scales in proportion to the **square** of the increase in one dimension and the volume scales in proportion to the **cube** of the increase in one dimension. In symbols:

area $\propto x^2$

volume $\propto x^3$

A quantity that depends on the area or the volume will also scale in the same way. So mass and weight (which are proportional to volume) will scale as the cube of the linear dimensions whereas rate of loss of thermal energy (which is proportional to the surface area) will scale as the square of the linear dimension.

The same principles can be applied to living objects. For example, we can consider the effect of scaling up a person by a factor of 1.5.

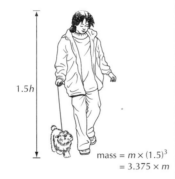

$1.5h$

h

mass = m

mass = $m \times (1.5)^3$
= $3.375 \times m$

The scaled-up version is 1.5 times taller than the original, but is more than three times the mass. Some useful quantities are given in the table below along with how they scale.

Absolute quantity	Scaling is proportional to...
Height	x
Surface area	x^2
Cross-sectional area	x^2
Volume	x^3
Mass	x^3
Weight	x^3
Muscle force	x^2 (the force $\propto$ the cross-sectional area of the muscle)

Since different quantities scale in different ways, scale models cannot behave in exactly the same way as the original objects.

ABSOLUTE AND RELATIVE PHYSICAL QUANTITIES

The previous box dealt with the procedure for working out the scaling of **absolute physical quantities** – in other words physical quantities that are directly related to the size of the object. For a living object it is often more important to consider the scaling of **relative physical quantities** – quantities that depend on several absolute physical quantities.

For example, it is important for animals to be able to control the rate of loss of thermal energy. This is an absolute physical quantity that scales as the square of the linear dimensions. But it is not just the quantity of energy that is important. If a small block and a large block were to lose the same amount of thermal energy, the small block's temperature would go down by a larger amount – the temperature change is proportional to the mass.

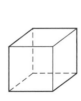
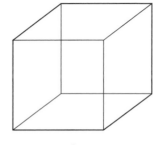

energy loss = 4200 J
temperature loss = 1 °C

energy loss = 4200 J
temperature loss = 0.125 °C

It is very important for small animals to control the loss of heat as even a small loss of energy can result in a large temperature change. This could affect the animal's ability to stay alive. The relative physical quantity that applies in this case is the rate of loss of temperature. Since this depends on both the rate of loss of heat and the amount of mass, we should consider the rate of loss of heat per unit mass.

Rate of loss of heat $\propto$ (surface area)

                         $\propto x^2$

Mass $\propto x^3$

So rate of loss of heat per unit mass

                 $\propto \left(\dfrac{x^2}{x^3}\right)$

                 $\propto x^{-1}$

The rate of loss of heat per unit mass (or the rate of loss of temperature) is inversely proportional to the liner dimensions. If the linear dimension were to halve, the rate of loss of temperature would double – and so on.

Some useful quantities are given in the table below along with how they scale.

Relative quantity	Scaling is proportional to...
Heat loss rate per unit mass	x^{-1}
Bone stress (force per unit area)	$\left(\dfrac{x^3}{x^2}\right)$ or x
Oxygen absorption rate per unit mass	$\left(\dfrac{x^2}{x^3}\right)$ or x^{-1}

Size, form and function

APPLICATION OF SCALING TO REAL SITUATIONS

As relative physical quantities scale in different ways, creatures of similar form (e.g. a dog and a horse) are not simply scaled versions of exactly the same shapes. Differences in size must also mean that the shape alters. Taken to extremes, larger or smaller versions of creatures are not always viable. The examples below show how the application of scaling and simple physical principles can be used to explain or predict the difference between creatures.

1. Why giants do not exist

Giants do not exist

When a person is standing up, the weight of the person means that the bones in their legs are under a certain amount of stress. Since stress is defined as force per unit cross-sectional area, it must scale as follows:

The force on either leg must be proportional to the weight of person. This scales as x^3. The cross-sectional area of leg, however, scales as x^2. This means that overall, the stress in the bone scales as x – it is proportional to the linear dimension. A giant that is ten times the height of a normal human would have ten times the stress in his bones just standing up. Normal bones subjected to this amount of stress would break.

2. Why the gazelle and the eland have different shaped legs

Both creatures are antelopes but the smaller gazelle has relatively slender legs when compared with the larger eland.

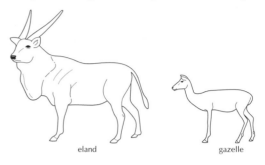

eland   gazelle

The reason for this is exactly the same as the previous example. If the eland was just a scaled version of the gazelle, then the stress in its legs would be too great. They have disproportionately thicker legs so as to be able to cope with this large stress.

3. Why aspects of the physical world change their significance at different sizes.

The pond skater is an insect that relies on the molecular forces on the surface of the liquid (the **surface tension**) in order to remain supported at the surface of a pond. Since weight scales as x^3 and surface tension scales as x^2, the weight of the insect soon becomes too great to be supported. No large versions of this insect exist.

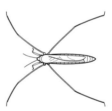

A pond skater relies on surface tension

Thus these molecular forces are significant for the small insect, but they would be irrelevant on a larger creature.

EXAMPLES

Given the information in the diagram, it is possible to predict the relative thickness of the leg bones for a pig and a rhinoceros.

We start by assuming that the bone stress is the same value for both creatures.

$$\frac{\text{length of pig}}{\text{length of rhinoceros}} = \frac{1}{3}$$

$$\text{therefore } \frac{\text{mass of pig}}{\text{mass of rhinoceros}} = \frac{1}{27}$$

$$\therefore \frac{\text{weight of pig}}{\text{weight of rhinoceros}} = \frac{1}{27}$$

$$\text{since stress} = \frac{\text{force}}{\text{area}}, \text{ if the leg stress is the same for each animal}$$

$$\frac{\text{cross-sectional area of leg of pig}}{\text{cross-sectional area of leg of rhinoceros}} = \frac{1}{27}$$

$$\frac{\text{thickness of leg of pig}}{\text{thickness of leg of rhinoceros}} = \sqrt{\left(\frac{1}{27}\right)} = \frac{1}{5.2}$$

3*l*

l

Sound intensity and the dB scale

SOUND INTENSITY

The **sound intensity** is the amount of energy that a sound wave brings to a unit area every second. The units of sound intensity are W m^{-2}.

It depends on the amplitude of the sound. A more intense sound (one that is louder) must have a larger amplitude.

Intensity ∝ (amplitude)^2

This relationship between intensity and amplitude is true for all waves.

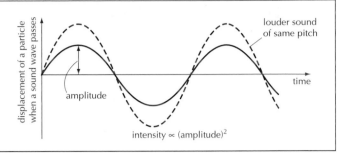

LOUDNESS

Intensity is a measurable quantity whereas loudness is subjective and depends on the listener.

Different people can describe different intensity sounds as appearing to have the same loudness – the frequency of the sound is an important factor. The following diagram shows the intensity – frequency diagram for a person with normal hearing.

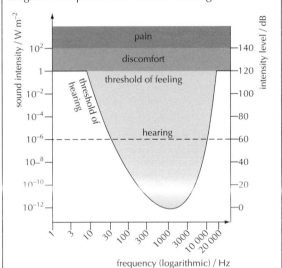

Diagram of intensity versus frequency for the average ear

The following points should be noted:
- the lines represent sounds that are perceived to be the same loudness.
- below the threshold of hearing line, nothing is audible.
- above the threshold of feeling line, the sound is becoming painful.
- the ear is most sensitive to sounds at a frequency of about 2 kHz.
- a standard frequency for measurement is 1kHz. At this frequency, the threshold of hearing corresponds to an amplitude of 10^{-12} W m^{-2}.
- this diagram represents the response of a normal ear. Medical problems or simple ageing would alter this diagram.

DECIBEL SCALE

Sound intensity levels are measured on the decibel scale (dB). As its name suggests, the decibel unit is simply one tenth of a base unit that is called the bel (B). Human hearing can respond to a huge range of different sound intensities. The decibel scale is logarithmic. The scale compares any given sound intensity with intensity at the threshold of hearing (the weakest sound that a person can normally hear). This threshold value is taken to be exactly 1 picowatt per square metre or 10^{-12} W m^{-2}.

Mathematically

$$\text{Intensity level in bels} = \log_{10} \frac{\text{intensity of sound}}{\text{intensity at the threshold of hearing}}$$

$$\text{or Intensity level in bels} = \log_{10} \frac{I}{I_0} \text{ (where } I_0 = 10^{-12} \text{ W m}^{-2})$$

Since 1 bel = 10 dB

$$\text{Intensity level in dB} = \log_{10} \frac{I}{I_0}$$

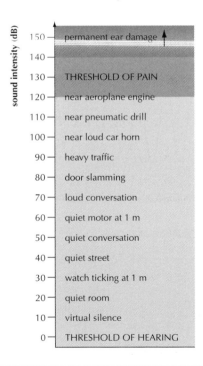

EXAMPLE

The intensity level next to an aircraft that is taking off is about 1 W m^{-2}. This means that the sound intensity level is given by

$$I = 10 \log \left(\frac{1}{10^{-12}} \right) \text{ dB}$$
$$= 10 \log (10^{12}) \text{ dB}$$
$$= 10 \times 12 \text{ dB}$$
$$= 120 \text{ dB}$$

The ear and the mechanism of hearing

THE EAR

The ear converts sounds into electrical signals in the brain. There are three main sections – the **outer**, the **middle** and the **inner** ear.

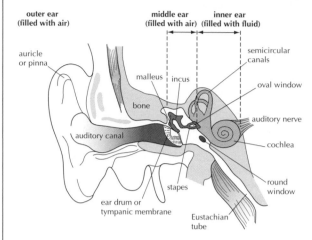

outer ear (filled with air) — middle ear (filled with air) — inner ear (filled with fluid)

auricle or pinna
malleus
incus
bone
semicircular canals
oval window
auditory nerve
auditory canal
cochlea
stapes
round window
ear drum or tympanic membrane
Eustachian tube

- The shape of the outer ear allows air vibrations to arrive at the ear drum (**tympanic membrane)**.
- The middle ear converts the oscillations of air (and so oscillations of the ear drum) into oscillations in the fluid in the inner ear at the **oval window**. Three small bones that are called the **malleus**, **incus** and **stapes** (or hammer, anvil and stirrup) achieve this. Collectively they are known as the **ossicles**.

- The inner ear (and in particular the **cochlea**) converts the oscillations in the fluid into electrical signals that are sent along the auditory nerve to the brain.

Part of the reason for the complexity is the need to transmit (rather than reflect) as much as possible of the sound from the air into the cochlear fluid. Technically achieving this transmission of energy is known as **impedance matching**.

- The arrangement of the ossicles is such that it achieves a **mechanical advantage** of about 1.5. This means that the bones act as levers and pistons and the forces on the eardrum are increased by 50% by the time they are transmitted to the oval window of the inner ear.
- The area of the oval window is about 15 times smaller than the area of the eardrum. This means that the pressure on the oval window will be greater than the pressure on the eardrum.

These two processes result in larger pressure variations in the cochlear fluid as compared to the pressure variations on the eardrum.

Other parts of the ear's structure include the **semicircular canals** and the **Eustachian tube**.

- The semicircular canals in the inner ear are not used for hearing sounds. They are involved in detecting movement and keeping the body balanced.
- The Eustachian tubes connect the middle ear to the mouth and allow the pressures on either side of the eardrum to be equalised. Although the tube is normally closed, it opens during swallowing, yawning or chewing.

PITCH

The pitch of a sound corresponds to the frequency of the wave – the higher the frequency of the sound, the higher the pitch.

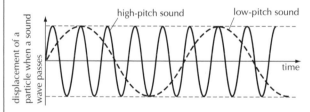

displacement of a particle when a sound wave passes — high-pitch sound — low-pitch sound — time

A normal human ear can hear sounds in the range 20 Hz up to 20 000 Hz (20 kHz).

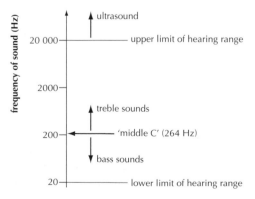

frequency of sound (Hz)

- ultrasound
- 20 000 — upper limit of hearing range
- 2000
- treble sounds
- 200 — 'middle C' (264 Hz)
- bass sounds
- 20 — lower limit of hearing range

Most sounds are not just one pure frequency – they are a mixture of different frequencies. The particular mixture used makes different sounds unique. The different frequencies present in the sound wave are distinguished in the cochlea.

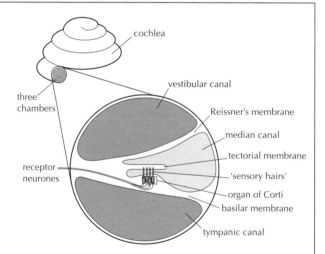

cochlea
vestibular canal
Reissner's membrane
three chambers
median canal
tectorial membrane
'sensory hairs'
organ of Corti
receptor neurones
basilar membrane
tympanic canal

The cochlea is a spiral that contains three chambers. The pressure wave starts from the oval window and passes to the top of the spiral along the **vestibular chamber**. It then returns via the **tympanic chamber** and ends up being absorbed at the round window. As the pressure wave travels along these two chambers, 'sensors' in the middle chamber (**cochlear duct**) convert the variations in pressure into electrical signals in the auditory nerve. The operation of these 'sensors' is not fully understood, but they involve small hair-like structures in something called the **organ of Corti**. As these hair-like structures are moved back and forth, electrical impulses are sent to the brain. The hair-cells are set in motion by movements in the **basilar membrane**. This consists of fibres that change in length and tension as one travels along the chambers. Different frequencies are detected by the different sized structures along the cochlea.

Hearing tests

EXPERIMENTAL PROCEDURES

The aim of a basic hearing test is to see how hearing ability varies with frequency. To gain a rough idea of how a person's hearing responds to frequency it is possible to use tuning forks as the source of known frequencies. If the hearing is poor in a particular frequency range, then a tuning fork would need to be brought closer to the ear before it was heard.

For proper diagnosis, a complete record of the variation of hearing with frequency is required. This is called an **audiogram**. A complete audiogram tests for both **air conduction** and **bone conduction** as a comparison between the two can help to find out what part of the hearing mechanism has gone wrong. In each case, the patient is placed in a soundproofed room and the quietest sounds that can be heard is recorded at a given frequency. The test is then repeated at different frequencies and then for the other ear.

- Air conduction is usually tested using headphones. At any one given frequency, sounds of varying intensity are fed into the ear and the threshold of hearing is established. This process is repeated at different frequencies.

- Bone conduction is tested by placing a vibrator on the bone behind the ear. It is held in place by a small band stretching over the top of the head. The vibrations are carried through the bones and tissues of the skull directly to the cochlea. This process allows the tester to bypass the outer ear and the middle ear. It tests the sensitivity of the inner ear directly.

In both tests, the amount of hearing loss is measured at different frequencies. A typical set of data would include the following frequencies:

125, 250, 500, 1000, 2000, 4000 and 8000 Hz

Before the start of the test, 'normal' levels of hearing have been established for the different frequencies used. The difference between this normal level and the recorded level is what is recorded on the audiogram using the dB scale. Different symbols are used to represent air conduction and bone conduction for the left and for the right ear. Air conduction thresholds are often represented by circles or crosses and bone conduction by triangles

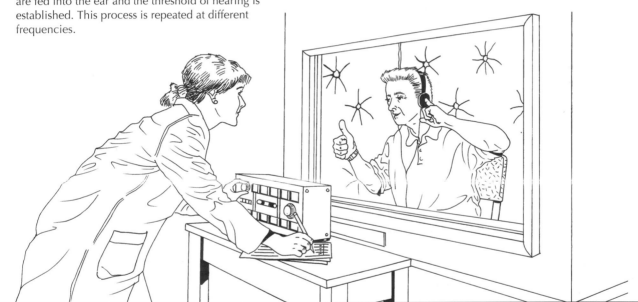

AUDIOGRAMS

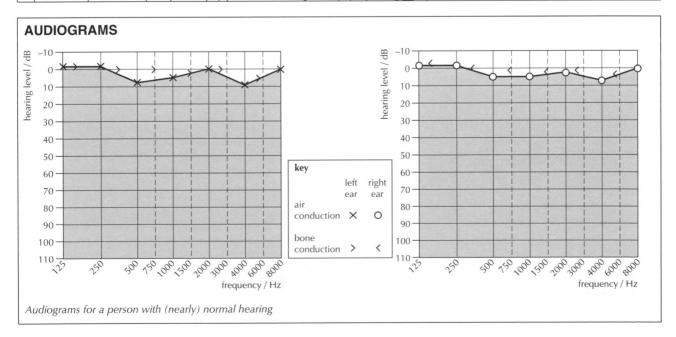

Audiograms for a person with (nearly) normal hearing

Hearing defects and correction

CONDUCTIVE LOSS

If the air conduction thresholds show a hearing loss but the bone conduction thresholds are normal, this is called a **conductive loss** of hearing. The sounds are being processed correctly in the inner ear, but the vibrations are not reaching it. This can sometimes be corrected by surgery. The main causes for conductive losses are:

- blockages – build up of wax or fluid.
- accidents – the eardrum can be ruptured or the middle ear could be damaged.
- diseases – the bones in the middle ear (and the oval window) can be prevented from moving.
- age – with increasing age, the bones in the middle ear (and the oval window) tend to become solidified.

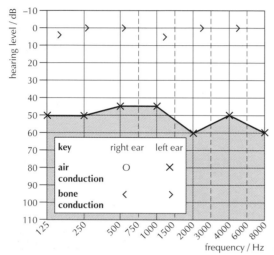

Audiogram for conductive loss (in left ear)

SENSORY LOSS

If both air conduction and bone conduction thresholds show the same amount of hearing loss, we call it **sensory** (**sensorimeural**) hearing loss. It is also possible to have a greater hearing loss in air conduction as compared to bone conduction. This would be called a **mixed** hearing loss.

Sensory loss can be caused by ageing or the exposure to excessive noise over periods of time.

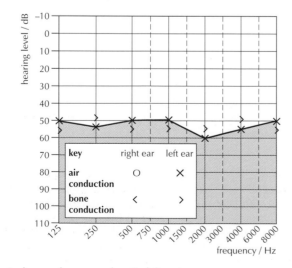

Audiogram for sensory loss (in left ear)

SELECTIVE FREQUENCY LOSS

All of the previous audiograms show a similar hearing loss at all frequencies. The audiogram below shows conductive loss particularly in the low and mid frequency range. This could lead to loss in speech discrimination.

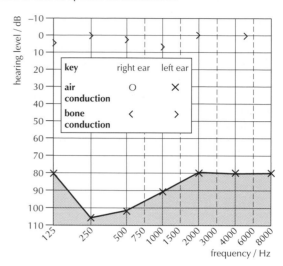

Audiogram for selective frequency loss (in left ear)

HEARING AIDS

Hearing aids attempt to correct for hearing loss. Three basic possible aids to hearing include **analogue aids**, **digital aids** and **cochlear implants**.

- There are many different types and sizes of analogue hearing aid, but they all do the same thing. They contain a microphone that converts the sound signal into an electrical signal. This electrical signal can be filtered and amplified before arriving at a miniature loudspeaker. A small battery powers the hearing aid. Some hearing aids are equipped with a small coil to receive signals from, for example, the television or a sound system installed in a theatre. Although simple and cheap, they tend to compress the audio range available. Some element of background noise will also be introduced.

- Digital hearing aids are the same as analogue hearing aids but the sound processing is digital. Thus, it is possible to refine the sound signal, for example by reducing noise and improving speech signals, or by amplifying only the frequencies that the user needs amplified. The digital circuit itself has no internal noise. This greater flexibility comes at a greater level of complexity and thus cost.

- Both analogue and digital hearing aids are of no use to somebody with profound sensory hearing loss. Cochlear implants bypass the outer and middle ears by using electrical stimulation of electrodes implanted in the cochlea. This involves surgery to place the electrodes inside the cochlea. A microphone in a behind-the-ear hearing aid case is connected to a package of electronics that is worn on a belt or carried in a pocket. The electronics translates the microphone signal into a set of four to eight electrical stimuli. These are sent to an array of electrodes implanted in the deaf patient's cochlea. These electrodes stimulate the nerves and the brain interprets this as sound.

X-ray imaging

INTENSITY, QUALITY AND ATTENUATION

The details of X-ray production are shown on page 84. The effects of X-rays on matter depend on two things, the **intensity** and the **quality** of the X-rays.

- The intensity, I, is the amount of energy per unit area that is carried by the X-rays.
- The quality of the X-ray beam is the name given to the spread of wavelengths that are present in the beam. The incoming energy of the X-rays arrives in 'packets' of energy called photons (see page 81 for more details). A typical X-ray spectrum is shown on page 84.

If the energy of the beam is absorbed, then it is said to be **attenuated**. If there is nothing in the way of an X-ray beam, it will still be attenuated as the beam spreads out. Two processes of attenuation by matter, **simple scattering** and the **photoelectric effect** are the dominant ones for low-energy X-rays.

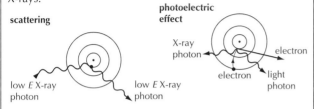

- Simple scattering affects X-ray photons that have energies between zero and 30 keV.
- In the photoelectric effect, the incoming X-ray has enough energy to cause one of the inner electrons to be ejected from the atom. It will result in one of the outer electrons 'falling down' into this energy level. As it does so, it releases some light energy. This process affects X-ray photons that have energies between zero and 100 keV.

Both attenuation processes result in a near exponential transmission of radiation as shown in the diagram below. For a given energy of X-rays and given material there will be a certain thickness that reduces the intensity of the X-ray by 50%. This is known as the **half-value thickness**.

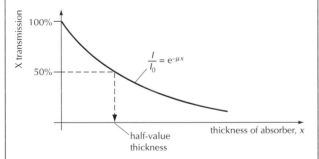

The **attenuation coefficient** μ is a constant that mathematically allows us to calculate the intensity of the X-rays given any thickness of material. The equation is as follows:

$$I = I_0\, e^{-\mu x}$$

The relationship between the attenuation coefficient and the half-value thickness is

$$x_{\frac{1}{2}} = \frac{\ln 2}{\mu}$$

$x_{\frac{1}{2}}$ The half-value thickness of the material (in m)

ln 2   The natural log of 2. This is the number 0.6931

μ The attenuation coefficient (in m^{-1})

BASIC X-RAY DISPLAY TECHNIQUES

The basic principle of X-ray imaging is that some body parts (for example bones) will attenuate the X-ray beam much more than other body parts (for example skin and tissue). Photographic film darkens when a beam of X-rays are shone on them so bones show up as white areas on an X-ray picture.

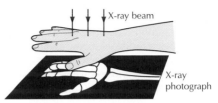

Since X-rays cause ionisations, they are dangerous. This means that the intensity used needs to be kept to an absolute minimum. This can be done by introducing something to **intensify** (to enhance) the image. There are two simple techniques of **enhancement**:

- When X-rays strike an intensifying screen the energy is re-radiated as visible light. The photographic film can absorb this extra light. The overall effect is to darken the image in the areas where X-rays are still present.
- In an image-intensifier tube, the X-rays strike a fluorescent screen and produce light. This light causes electrons to be emitted from a photocathode. These electrons are then accelerated towards an anode where they strike another fluorescent screen and give off light to produce an image.

X-RAY IMAGING TECHNIQUES INCLUDING COMPUTER TOMOGRAPHY

With a simple X-ray photograph it is hard to identify problems within soft tissue, for example in the gut. There are two general techniques aimed at improving this situation.

- In a **barium meal**, a dense substance is swallowed and its progress along the gut can be monitored.
- **Tomography** is a technique that makes the X-ray photograph focus on a certain region or 'slice' through the patient. All other regions are blurred out of focus. This is achieved by moving the source of X-rays and the film together.

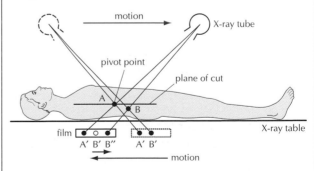

An extension of basic tomography is the **computerised tomography scan** or **CT scan**. In this set-up a tube sends out a pulse of X-rays and a set of sensitive detectors collects information about the levels of X-radiation reaching each detector. The X-ray source and the detectors are then rotated around a patient and the process is repeated. A computer can analyse the information recorded and is able to reconstruct a 3-dimensional 'map' of the inside of the body in terms of X-ray attenuation.

Ultrasonic imaging

ULTRASOUND AND SONAR

Ultrasound

The limit of human hearing is about 20 kHz. Any sound that is of higher frequency than this is known as **ultrasound**. Typically ultrasound used in medical imaging is just within the MHz range. The velocity of sound through soft tissue is approximately 1500 m s^{-1} meaning that typical wavelengths used are of the order of a few millimetres.

Unlike X-rays, ultrasound is not ionising so it can be used very safely for imaging inside the body – with pregnant women for example. The basic principle is to use a probe that is capable of emitting and receiving pulses of ultrasound. The ultrasound is reflected at any boundary between different types of tissue. The time taken for these reflections allows us to work out where the boundaries must be located.

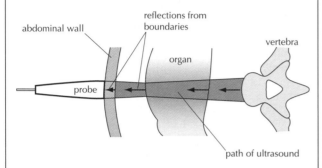

Very strong reflections take place when the boundary is between two substances that have very different densities. This can cause some difficulties.

- In order for the ultrasound to enter the body in the first place, there needs to be no air gap between the probe and the patient's skin. An air gap would cause almost all of the ultrasound to be reflected straight back. The transmission of ultrasound is achieved by putting a gel or oil (of similar density to the density of tissue) between the probe and the skin.
- Very dense objects (such as bones) can cause a strong reflection and multiple images can be created. These need to be recognised and eliminated.

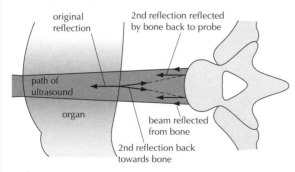

Sonar

The nature and position of the boundaries between organs and tissues are worked out using the same principles by which a ship can locate objects underwater. This technique is called **SONAR** or 'Sound Navigation and Ranging'. A pulse of ultrasound is sent into the body. When an echo returns, it is converted back into an electrical signal and thus the time for the return journey of the ultrasound can be calculated. From this time taken, the distance to the boundary can be worked out.

A- AND B-SCANS

There are two ways of presenting the information gathered from an ultrasound probe, the **A-scan** or the **B-scan**. The A-scan (amplitude-modulated scan) presents the information as a graph of signal strength versus time. The B-scan (brightness-modulated scan) uses the signal strength to affect the brightness of a dot of light on a screen.

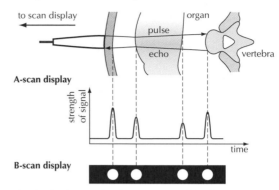

Pulse display by amplitude (A-scan) and brightness (B-scan)

A-scans are useful where the arrangement of the internal organs is well known and a precise measurement of distance is required. If several B-scans are taken of the same section of the body at one time, all the lines can be assembled into an image which represent a section through the body. This process can be achieved using a large number of transducers.

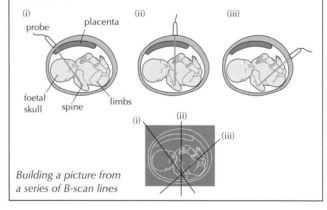

Building a picture from a series of B-scan lines

CHOICE OF FREQUENCY

The choice of frequency of ultrasound to use can be seen as the choice between resolution and attenuation.

- Here, the resolution means the size of the smallest object that can be imaged. Since ultrasound is a wave motion, diffraction effects will be taking place. In order to image a small object, we must use a small wavelength. If this was the only factor to be considered, the frequency chosen would be as large as possible.
- Unfortunately attenuation increases as the frequency of ultrasound increases. If very high frequency ultrasound is used, it will all be absorbed and none will be reflected back. If this was the only factor to be considered, the frequency chosen would be as small as possible.

On balance the frequency chosen has to be somewhere between the two extremes. It turns out that the best choice of frequency is often such that the part of the body being imaged is about 200 wavelengths of ultrasound away from the probe.

Other imaging techniques

NMR

Nuclear Magnetic Resonance (NMR) is a very complicated process but one that is extremely useful. It can provide detailed images of sections through the body without any invasive or dangerous techniques. It is of particular use in detecting tumours in the brain.

In outline, the process is as follows:
- the nuclei of atoms have a property called spin.
- the spin of these nuclei means that they can act like tiny magnets.
- these nuclei will tend to line up in a strong magnetic field.

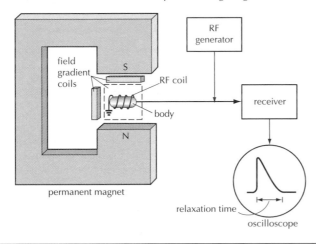

- they do not, however, perfectly line up – they oscillate in a particular way that is called **precession**. This happens at a very high frequency – the same as the frequency of radio waves.
- the particular frequency of precession depends on the magnetic field and the particular nucleus involved. It is called the **Larmor frequency**.
- if a pulse of radio waves is applied at the Larmor frequency, the nuclei can absorb this energy in a process called **resonance**.
- after the pulse, the nuclei return to their lower energy state by emitting radio waves.
- the time over which radio waves are emitted is called the **relaxation time**.
- the radio waves emitted and their relaxation times can be processed to produce the NMR scan image.

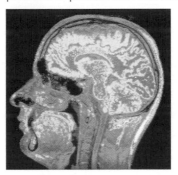

NMR scan image of the head

RADIOACTIVE TRACERS

Radioactive tracers can be used as 'tags'. If they are introduced into the body, their progress around the body can be monitored from outside (so long as they are gamma emitters). This can give information about how a specific organ is (or is not) functioning as well as being used to analyse a whole body system (for example the circulation of blood around the body).

There are many factors that affect the choice of a particular radioisotope for a particular situation. Some of these are listed below:
- the radioisotope should be able to be 'taken up' by the organ in question in its usual way. In other words it needs to have specific chemical properties.
- the quantity of radioisotope needs to be as small as possible so as to minimise the harmful ionising radiation received by the body.
- the lifetime of the tracer needs to be matched to the time scale of the process being studied.

Organ/tissue	Tracers	Uses
general body composition	^{3}H, ^{24}Na, ^{42}K, ^{82}Br	Used to measure volumes of body fluids and estimate quantities of salts (e.g. of sodium, potassium, chlorine).
blood	^{32}P, ^{51}Cr, ^{125}I, ^{131}I, ^{132}I	Used to measure volumes of blood and the different components of blood (plasma, red blood cells) and the volumes of blood in different organs. Also used to locate internal bleeding sites.
bone	^{45}Ca, ^{47}Ca, ^{85}Sr, ^{99m}Tc	Used to investigate absorption of calcium, location of bone disease and how bone metabolizes minerals.
cancerous tumours	^{32}P, ^{60}Co, ^{99m}Tc, ^{131}I	Used to detect, locate, and diagnose tumours. ^{60}Co is used to treat tumours.
heart and lungs	^{99m}Tc, ^{131}I, ^{133}Xe	Used to measure cardiac action: blood flow, volume, and circulation. Labelled gases used in investigations of respiratory activity.
liver	^{32}P, ^{99m}Tc, ^{131}I, ^{198}Au	Used in diagnosing liver disease and disorders in hepatic circulation.
muscle	^{201}Tl	Diagnosis in organs; in particular, heart muscle.

PET SCANS

Positron Emission Tomography (PET) is a technique that uses an isotope of carbon, carbon-11, as the radioactive tracer to measure the changes in blood flow within the brain. This particular isotope can be introduced into the body by breathing in a small sample of carbon monoxide. The radioactive carbon atoms will be easily taken up by the red blood cells and flow around the body.

When the carbon-11 decays, it does so by positive beta decay. A positron is emitted. When the positron meets an electron, they will destroy each other and two gamma rays will be emitted. These two gamma rays will essentially be travelling in opposite directions. Detectors around the patient can pick up these two gamma rays and calculate with great accuracy the position of the source of the gamma rays.

Current researchers can use this technique to image the functioning of the brain. A patient is given a specific task (such as reading or drawing) and the PET scan can identify the areas of the brain that receive more blood as a result of doing the task.

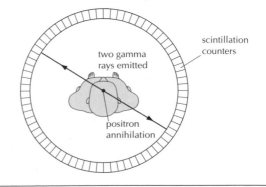

CENTRE OF GRAVITY

The **centre of gravity** of an object is the point where the net force of gravitational attraction appears to act. In principle, every part of an object feels a separate gravitational attraction towards the Earth – the overall effect is as though there was one much larger force acting at the centre of gravity.

The centres of gravity for non-symmetrical objects are relatively easy to estimate.

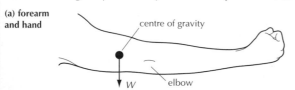

EQUILIBRIUM

An object is in equilibrium if it is at rest or if it is moving at constant velocity.

For example, if you bend your elbow and keep your forearm horizontal, then the forearm must be in equilibrium while it is stationary. There must be no resultant force and no resultant moment.

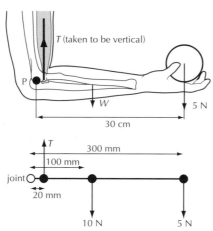

The addition of the object means that

$$T \times 2\text{ cm} = (W \times 10\text{ cm}) + (5\text{ N} \times 30\text{ cm})$$
$$= 100 + 150\text{ N cm} = 250\text{ N cm}$$
$$\therefore \quad T = \frac{250\text{ N cm}}{2\text{ cm}} = 125\text{ N}$$

It should be noted that in the above example, the addition of the 5N object caused the forces in the muscles and at the joints to increase by a lot more than 5N. (Without object $T = 50$N)

This is the underlying reason why it is important to try and avoid lifting a heavy object with a bent back. The forces at the base of the spine are large when the back is bent. If an additional load is carried at the same time, the total force can be very large indeed.

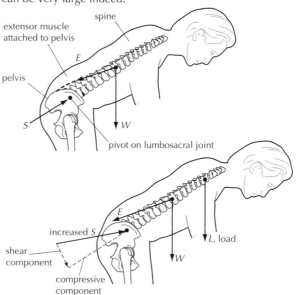

MECHANICAL ADVANTAGE AND VELOCITY RATIO

Many parts of the body can be considered to be a type of lever if they involve a pivot at a joint. In general muscles will apply a force (the **effort**). This will result in a turning around the pivot and another force (the **load**) being supported or balanced. The ratio between load and effort is the **mechanical advantage**.

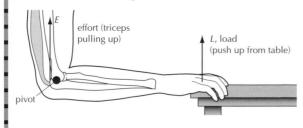

Limb structures in the body tend to have a mechanical advantage less than 1 – a mechanical disadvantage.

There is, however, a corresponding movement advantage in having things this way. The movement advantage is measured by a quantity called the **velocity ratio** (movement ratio) for the system. It does not have any units. If the mechanical advantage of a limb structure is less than one then the corresponding velocity ratio would be expected to be greater than one.

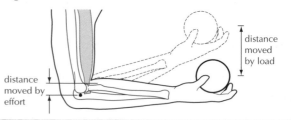

METABOLIC RATE

The **metabolic rate** is the rate at which energy is being converted in an organism. This includes even the involuntary functions such as the beating of the heart or breathing. The rate of using energy for all these involuntary functions (including such things as digestion and growth) is called the **basal metabolic rate (BMR)**.

- BMR tends to decrease with age. Initially a lot of energy is used in the growing process. As we get older, all processes tend to slow down.
- after childhood, BMR is generally higher in males than in females. This is to do with the different proportions of fat in the different sexes. On average, females have a higher proportion than males.

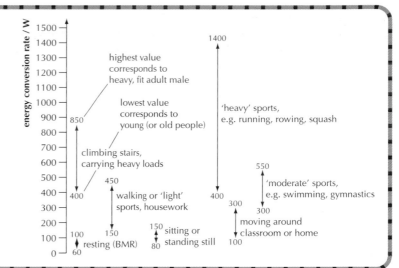

TEMPERATURE REGULATION

The optimum temperature for the human body to operate at is approximately 37 °C. A variation in core temperature of more than about 6 °C will result in death. If the temperature of the surroundings is greater than this then the body will have to expend energy in order to cool down. If the temperature of the surroundings is less than this then the body will need to expend energy replacing thermal energy lost.

The main processes that result in the loss of thermal energy are convection, conduction, radiation, evaporation, respiration and expiration (breathing out).

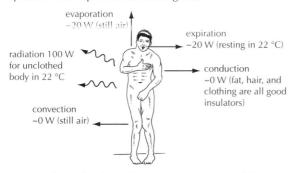

- **Convection.** This depends on the temperature difference between the body and the surroundings and on the presence of a draught. The rate of loss of heat via this process tends to be small.
- **Conduction.** This depends on the temperature difference between the body and the surroundings, the surface area

of the body and the thermal conductivities involved. In general, the rate of loss of heat via this process is very low because hair, fat and clothing are all good insulators.

- **Radiation.** This depends on the absolute temperature of the body, the surface area and the type of surface. Typically, it is in the region of 100 W.
- **Evaporation.** This depends on the temperature difference between the body and the surroundings, the surface area of the skin exposed, the humidity of the air (% water content) and on the presence of a draught.
- **Expiration.** When carbon dioxide and water vapour are expired, they take thermal energy away from the body.

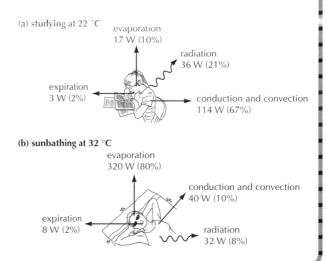

MATHEMATICAL EXAMPLES

The food we eat provides us with a certain amount of energy. The actual amount depends on the efficiency of the conversion process. We use up this energy doing tasks and staying alive. Different tasks will require different amounts of energy. If we are provided with more energy than we need, the body 'stores' it in the form of fat. If we use more energy than we take in, the body must use up its stores.

e.g. a bowl of soup might be able to provide 985 kJ energy.

The time taken to use this up is different.

If swimming, (rate of energy expenditure = 500 W)

$$\text{time taken before energy is used up} = \frac{985000}{500}$$

$$= 1970 \text{ s}$$

$$= 32.8 \text{ minutes}$$

If resting, (BMR = 75 W)

$$\text{time taken before energy is used up} = \frac{985000}{75}$$

$$= 13133 \text{ s}$$

$$= 3.6 \text{ hours}$$

Biological effects of radiation, dosimetry and radiation safety

EFFECTS OF RADIATION

At the molecular level, an ionisation could cause damage directly to a biologically important molecule such as DNA or RNA. This could cause it to cease functioning. Alternatively, an ionisation in the surrounding medium is enough to interfere with the complex chemical reactions (called **metabolic pathways**) taking place.

Molecular damage can result in a disruption to the functions that are taking place within the cells that make up the organism. As well as potentially causing the cell to die, this could just prevent cells from dividing and multiplying. On top of this, it could be the cause of the transformation of the cell into a malignant form.

As all body tissues are built up of cells, damage to these can result in damage to the body systems that have been affected. The non-functioning of these systems can result in death for the animal. If malignant cells continue to grow then this is called **cancer**.

DOSIMETRY

Three quantities are of particular note – **exposure, absorbed dose** and **dose equivalent**.

- **Exposure** is a measure of the total amount of ionisation produced. It is defined as

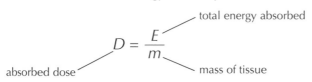

$$X = \frac{Q}{m}$$

exposure — total charge of one sign produced — mass of air

The units of exposure X are coulombs per kilogram ($C\ kg^{-1}$)

- The **absorbed dose** is the energy absorbed per unit mass of tissue.

$$D = \frac{E}{m}$$

absorbed dose — total energy absorbed — mass of tissue

The units of absorbed dose D are joules per kilogram ($J\ kg^{-1}$). This unit is called the gray, Gy.

- The **dose equivalent** is an attempt to measure the radiation damage that actually occurs in tissues.

The dose equivalent is defined as

$$H = QD$$

dose equivalent — quality factor — absorbed dose

The units of dose equivalent H are once again joules per kilogram ($J\ kg^{-1}$). In order to differentiate between this and the absorbed dose, the unit is called the sievert (Sv).

RADIATION SAFETY

There is no such thing as a safe dose of ionising radiation. Any hospital procedures that result in a patient receiving an extra dose (for example having an X-ray scan) should be justifiable in terms of the information received or the benefit it gives.

There are three main ways of protecting oneself from too large a dose. These can be summarised as follows:

- **Run away!**
  The simplest method of reducing the dose received is to increase the distance between you and the source. Only electromagnetic radiation can travel large distances and this follows an inverse square relationship with distance.

- **Don't waste time!**
  If you have to receive a dose, then it is important to keep the time of this exposure to a minimum.

- **If you can't run away, hide behind something!**
  Shielding can always be used to reduce the dose received. Lead-lined aprons can also be used to limit the exposure for both patient and operator.

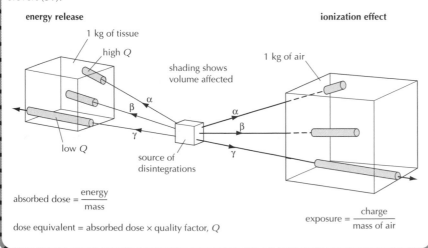

energy release

ionization effect

1 kg of tissue

high Q

shading shows volume affected

1 kg of air

β   α

α

β

γ

γ

low Q

source of disintegrations

$$\text{absorbed dose} = \frac{\text{energy}}{\text{mass}}$$

dose equivalent = absorbed dose × quality factor, Q

$$\text{exposure} = \frac{\text{charge}}{\text{mass of air}}$$

Radiation sources in diagnosis and therapy

DIFFERENT TYPES OF HALF-LIFE

A radioactive source will decay in the laboratory. If the source is introduced into a living body, there is the extra complication of the biological processes that are taking place. These will tend to remove the source from the body and this means that the **effective half-life** is reduced.

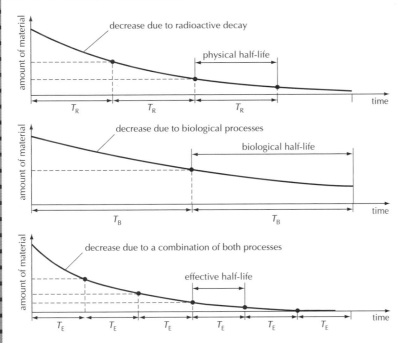

The effective half-life T_E of an isotope is the time taken for the number of radioactive nuclei of that isotope **in the body** to halve. This can be seen as the result of two processes. One process is radioactive decay – this will have a certain **physical half-life** T_R. The other process is the chemical removal of the isotope from the body and this will have a certain **biological half-life** T_B. The relationship between these values is

$$\frac{1}{T_E} = \frac{1}{T_R} + \frac{1}{T_B}$$

RADIATION THERAPY

So far, we have considered only the applications of ionising radiation involved with diagnostic medicine (identifying diseases), but they can also be used for therapy (treatment). The aim of radiotherapy is to target malignant cells in preference to normal healthy cells.

The dose that is used is critical. Too high a dose and too many healthy cells are killed. Too low a dose and the cancer is not destroyed. The dose needs to be as high as possible in the region of the cancer and as low as possible elsewhere. This can be achieved by one of two techniques:

- A radioactive source can be placed in the tumour itself. This can be done chemically or physically.
- Overlapping beams of radiation can be used. Where they overlap the dose will be high, elsewhere the dose will be lower.

The source of the ionising radiation can be a radioactive element, but these days it is also common to use high-energy X-rays or gamma rays from particle accelerators. High-energy protons can also be used directly.

CHOICE OF RADIOISOTOPE

Important factors include:

- the nature of the radiation – alpha, beta and gamma have different penetrating powers.
- the energy of the radiation.
- the physical half-life of the radioisotope. A short half-life means that the radioisotope will decay quickly, but sometimes a constant source is required.
- the biological half-life.
- the chemical properties – some chemicals are 'taken up' more readily than others.
- wether the radioisotope is solid, liquid or gas.
- the availability of the radioisotope.

Some common radioisotopes are listed below:

Radioisotope	Comment
Cobalt-60	Source of high-energy gamma rays used for radiotherapy. Long half-life so does not need to be replaced often.
Technetium-99m	This is a gamma emitter with a short half-life that can be produced relatively easily. It can be used in different forms to study blood flow, liver function, and bone growth.
Iodine-123	A gamma emitter that is readily taken up by the thyroid. It can also be used to study the functioning of the liver.
Iodine-131	A gamma emitter than is sometimes used to treat cancer of the thyroid.
Xenon-133	A gamma emitter in gaseous form which can be inhaled and used to study lung function.

EXAMPLE

If the physical half-life of a radioisotope is 10 days and the biological half-life is 15 days, what fraction will remain after 30 days?

$$\frac{1}{T_E} = \frac{1}{10} + \frac{1}{15} = \frac{1}{6}$$

$$\therefore T_E = 6$$

$$30 \text{ days} = 5 \times T_E$$

$$\therefore \text{ fraction remaining} = \left(\frac{1}{2}\right)^5 = 3.1\%$$

IB QUESTIONS – OPTION D – BIOMEDICAL PHYSICS

1 The diagram below shows how the typical threshold of hearing varies with frequency for a normal young person.

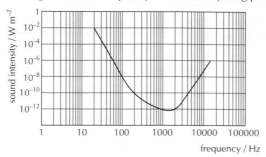

(a) Outline how the data for this graph could be obtained. [3]

(b) To what approximate frequency of sound is the ear most sensitive? [1]

(c) Over what range of frequencies is a sound of intensity 10^{-10} W m^{-2} audible? [1]

2 The diagram below shows a simplified model of a forearm, alongside a cm scale. The forearm is horizontal.

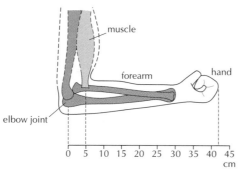

(a) Estimate the position of the centre of mass of this arm-hand system and mark this point with an X on the diagram. Explain your reasoning. [2]

(b) The arm-hand system has a total mass of 1.7 kg. Determine the torque about the elbow joint produced by the weight of the arm-hand system. [2]

(c) Determine the tension in the muscle necessary to support the arm in this position. [2]

3 This question is about scaling in mammals.

(a) Explain, by making suitable estimates, why, under the same external conditions, a baby is at a much greater risk of dying from exposure to cold than an adult. [4]

(b) State **one** assumption that you have made in the above explanation. [1]

4 This question is about human hearing.

(a) Explain what is meant by *conductive* hearing loss. [1]

(b) A person with conductive hearing loss who uses an effective hearing aid will suffer little loss of hearing when taking part in a conversation. However, explain why such a person should be advised **not** to spend a lot of money on a 'Hi-Fi' music system. [3]

(c) A person has hearing loss of 40 dB at a frequency of 1000 Hz. Estimate the least intensity of sound that the person can detect at this frequency. [2]

5 When a person lifts a suitcase, the spine experiences large extra forces. In a simplified model of the situation, the spine can be treated as a rigid rod.

In this model, when the suitcase is lifted, three extra forces act on the spine which need to be in equilibrium.

- The additional force due to lifting the suitcase, *S*.
- The additional force from the muscles, *F*.
- The additional force on the base of the spine, *R*.

The diagram below shows the directions and points of action of *S* and *F*, but not *R*.

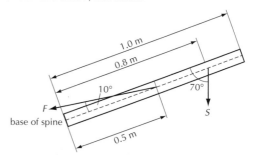

(a) State the **two** conditions for *S*, *F* and *R* to be in equilibrium. [2]

(b) Add an arrow to the diagram to show the approximate direction of *R*, the additional force on the base of the spine.

(c) Write down an expression for the torque about the base of the spine due to the force *S*. [2]

(d) Show that the force *F* is approximately nine times the force *S*, i.e. the muscle force is nine times the weight of the suitcase being lifted. [2]

6 Radioisotopes can be introduced into the body for **imaging** or for **therapy**. One common radioisotope is iodine-131.

(a) Explain the difference between **biological** half-life and **physical** half-life of a radioisotope. [2]

(b) A sample of a compound of Iodine-131 is administered to a patient. The physical half-life of Iodine-131 is 8 days whereas the biological half-life of this compound is about 20 days. What percentage activity will remain after 40 days? [3]

(c) If a patient receives the same **absorbed dose** from two different sources they would not necessarily receive the same **dose equivalent**. Explain what is meant by the terms **absorbed dose** and **dose equivalent**. Explain how they are related. [3]

(d) Outline **two** precautions necessary when introducing radioisotopes into the body. [2]

Astronomical observations

DURING ONE DAY

The most important observation is that the pattern of the stars remains the same from one night to the next. Patterns of stars have been identified and 88 different regions of the sky have been labelled as the different **constellations**. A particular pattern is not always in the same place, however. The constellations appear to move over the period of one night. They appear to rotate around one direction. In the Northern Hemisphere everything seems to rotate about the pole star.

It is common to refer measurements to the 'fixed stars' the patterns of the constellations. The fixed background of stars always appears to rotate around the pole star. During the night, some stars rise above the horizon and some stars set beneath it.

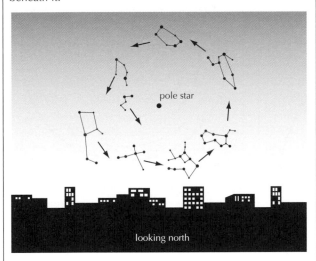

The same movement is continued during the day. The sun rises in the East and sets in the West, reaching its maximum height at midday. At this time in the Northern Hemisphere the Sun is in a southerly direction.

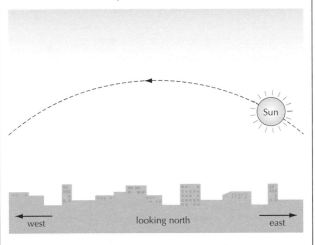

DURING THE YEAR

Every night, the constellations have the same relative positions to each other, but the location of the pole star (and thus the portion of the night sky that is visible above the horizon) changes slightly from night to night. Over the period of a year this slow change returns back to the exactly the same position

The Sun continues to rise in the East and set in the West, but as the year goes from winter into summer, the arc gets bigger and the Sun climbs higher in the sky.

FROM PLACE TO PLACE

If you move from place to place around the Earth, the section of the night sky that is visible over a year changes with latitude. The total pattern of the constellations is always the same, but you will see different sections of the pattern.

PLANETS

There are some objects in the night sky that do not exactly follow the pattern of the stars that has been described above. They do 'join in' with the nightly rotation, but from one night to the next, their positions change with respect to the pattern of the constellations. These objects are seen to slowly 'wander' around the night sky. The pattern of wandering is not fixed over one year. These objects are called 'the planets' which is derived from the Greek word for wanderer.

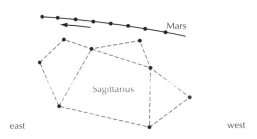

A particular aspect of the motion of the planets is the **retrograde motion** that is sometimes observed. Although their motion over several nights is usually a slow arc across the sky, they can sometimes seem to reverse their direction of travel as shown below.

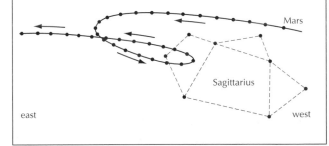

THE MOON

The Moon's observed motion through the night sky is different again. Although it also takes part in the nightly rotation around the pole star, it also wanders. On top of this it changes its appearance over the course of one month – it is said to have **phases**.

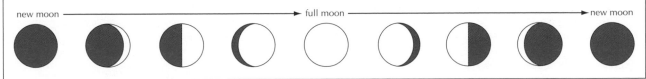

Development of models of the Universe

ARISTOTLE AND PTOLEMY

The geocentric ('Earth centred') model of the Universe has always been associated with the Greek philosopher Aristotle (384 BC – 322 BC). The Earth was at the centre of the Universe with the Heavens above. Each of the Heavenly objects (the Sun, Moon and planets) were attached to different concentric perfect crystal spheres. All the fixed stars were attached to the final sphere.

The spheres were turning all the time carrying the Heavenly objects with them. The spheres had to be moving at slightly different speeds.

Ptolemy (AD 85 – ~ 165) adapted the model. In order to explain the retrograde motion of some of the planets, he introduced the ideas of 'wheels within wheels'. A simple version allows each planet to move in a small circle called an **epicycle**. The centre of the epicycle moved on a bigger circle called the **deferent**.

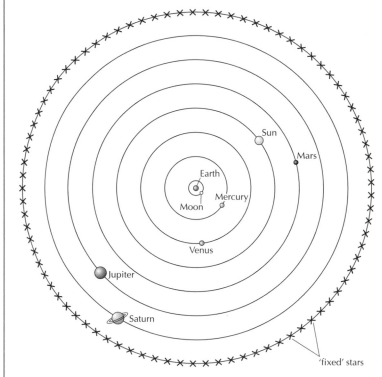

Aristotelian model without epicycles

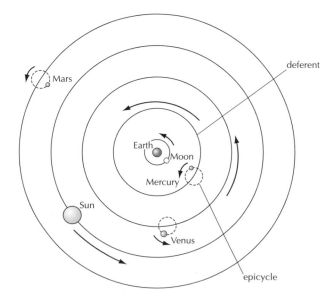

Epicycles used to explain retrograde motion

ARISTARCHUS AND COPERNICUS

An alternative model was also developed which was heliocentric ('sun centred'). Although first proposed by Aristarchus in second century BC, it has come to be associated with Copernicus (1473 – 1543). This model placed the Sun at the centre and relegated the Earth to be just one of the planets that orbit the Sun. The orbits of the planets were still taken to be circular.

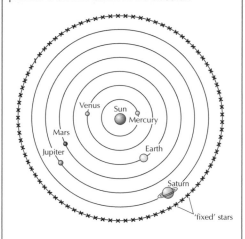

The great success of this model was that it could explain simply the retrograde motion of the planets.

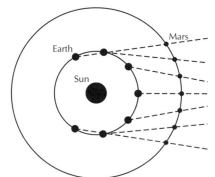

The direction in which Mars is seen from Earth varies as the planets move round the Sun; this accounts for the observed retrograde motion.

It was also able to explain why Mercury and Venus always appear close to the Sun in the sky. The disadvantage was that there were small but significant differences between the planets' motions as predicted by this model and practical observations. In order to account for this, Copernicus added epicycles to account for the slight variations.

Copernicus was aware that the heliocentric model was in conflict with the religious teachings of the Church at the time. His ideas were not widely accepted until they were popularised by Galileo in the seventeenth century.

Galileo, Kepler and Newton's synthesis

GALILEO

Galileo Galilei (1564 – 1642) added further support to the Copernican model of the Universe. As the first astronomer to construct a telescope, he was allowed a unique view of the heavens. He observed that
- there were mountains on the moon, just like the Earth.
- the planet Venus has phases (like the Moon) and it appeared to change in size.
- the planet Jupiter appeared to have other objects orbiting around it.
- there were dark spots on the surface of the Sun.
- the Sun rotated and the spots moved with its rotation

It was an attempt to publish his ideas and observations that finally brought him into direct conflict with the Roman Catholic Church. In 1633 he was tried by the Inquisition and forced to recant his ideas.

KEPLER

At about the same time as Galileo was supporting the Copernican model of the Universe, Johannes Kepler (1571–1642) was developing an adaptation of the basic model that gave extremely good agreement between prediction and observation.

Kepler was able to deduce three mathematical laws about the movement of the planets from Brahe's precise records. They are:
- The orbit of each planet around the Sun is an ellipse with the Sun at one focus.

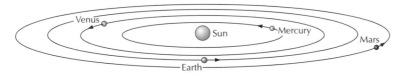

Planets travel around the Sun in ellipitical orbits, in the same direction and in roughly the same plane.

- A line joining a planet and the Sun sweeps out equal areas in equal time intervals.

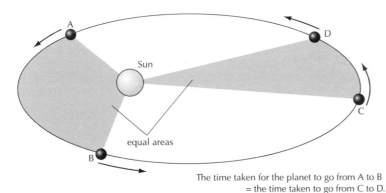

The time taken for the planet to go from A to B = the time taken to go from C to D.

- The square of a planet's orbital period is proportional to the cube of its average distance from the Sun.

It should be noted that Kepler's laws (particularly the third law) are mathematical observations rather than explanations in terms of physical principles, i.e. they are empirical laws.

UNIVERSAL GRAVITATION

It was left to the genius of Isaac Newton (1642 – 1727) to come up with a physical description that could explain Kepler's mathematical relationships. It is summarised in his Universal law of gravitation. The details of this law (and the derivation of Kepler's third law) are covered in the Higher Level course – see pages 61 and 63.

In this law, Newton proposes that every object in the Universe attracts every other object in the Universe with a gravitational force. The magnitude and direction of this force can be predicted from the masses involved and the distances that they are apart.

An important aspect of his explanation is that his law of gravitation is a 'universal law' – it applies everywhere in the Universe. In other words the law can explain how an apple falls down to the Earth but it can also explain how the Moon stays in orbit around the Earth. The same principle is used for these two very different situations.

gravitational force on apple from Earth

gravitational force on Moon from Earth

The gravitational force explains why an apple falls to the ground and also why the Moon stays in orbit.

KEPLER'S LAWS

Newton's law of gravitation is not just an explanation of the ideas of gravitational attraction, it makes a mathematical prediction about the magnitude of the force. Central to this law is the dependence on distance apart. The force is what is known as an 'inverse square' relationship. The force F between two masses is proportional to the inverse of the square of the distance r between the two masses.

$$F \propto \frac{1}{r^2}$$

It can be shown that a force of this general nature will result in an object taking up an elliptical orbit around a large mass. On top of this, it is possible to show directly that Kepler's third law can be derived from Newton's law of gravitation and his laws of motion. See page 63.

Development of concepts of motion, force and mass

ARISTOTLE

When analysing motion, Aristotle made the distinction between **natural motion** and **violent motion** (or forced motion). Everything was made up of a combination of four substances: Earth, Water, Fire and Air. The natural motions of Earth and Water were down whereas the natural motion of Fire and Air was up.

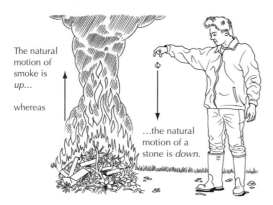

The natural motion of smoke is *up*...

whereas

...the natural motion of a stone is *down*.

He made two statements about natural motion:
- heavier things fall faster, the speed being proportional to the weight.
- the speed of fall of a given object depends *inversely* on the density of the medium it is falling through.

From the second statement, he concluded that a vacuum could not exist, because if it did, since it has zero density, all bodies would fall through it at infinite speed which clearly does not make sense.

For violent motion, Aristotle stated that the speed of the moving object was proportional to the applied force. This means that if you stop pushing an object, the object should stop moving. This is correct for many everyday situations – for example pushing a heavy block along a surface. But if you throw a stone horizontally, it does not fall to the ground (its natural motion) immediately after it has left your hand – why?

The Aristotelian explanation of the stone's motion would, of course, consider this to be an example of a violent motion. When the violent motion runs out, the natural motion takes over and, of course, the stone does end up on the ground.

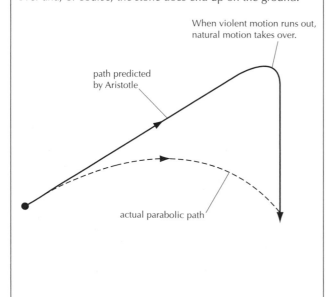

When violent motion runs out, natural motion takes over.

path predicted by Aristotle

actual parabolic path

GALILEO

In some senses, Galileo was responsible for the birth of the modern scientific method. The reliance on careful observation of experimental situations allowed him to propose theories that others could test. This is very different to Aristotle's approach to a problem. He tended to avoid experiments and was satisfied with developing a theory that could be justified and argued in principle. It is true that many of Aristotle's initial observations were experimental but he did not go on to test his assertions in practice.

Another new approach that Galileo brought the problem of motion was his reliance on mathematics. Without this analytical detail in his investigations much of his work on accelerated motion could not have taken place.

NEWTON VS. ARISTOTLE

	Newton	Aristotle
Motion	No force is required for an object to be in motion.	A force is required for violent motion. Natural motion can take place without a force.
Force	A resultant force changes the motion of (accelerates) an object. The acceleration is proportional to the force.	A force causes an object to gain a speed. The speed is proportional to the force.
Mass	Objects with different masses will fall together (in the absence of friction).	Objects with different masses will fall at different speeds.

MECHANICAL DETERMINISM

One aspect of Newtonian mechanics is that it is completely **deterministic**.

In theory, so long as we know the exact masses and the velocities of any system of particles, we can determine precisely the future arrangement at any given time later. We do this by applying Newton's laws every time there is a collision.

In practice we may not be able to take the initial measurements accurately enough to predict successfully exactly what is going to happen, but this does not change the fact that the future arrangement of the particles is exactly determined by the starting conditions. If the starting conditions were exactly repeated, then the subsequent positions would be identical.

According to Newtonian mechanics, the current positions and velocities of the atoms inside your body exactly determine the future arrangement of the atoms. Of course there will be interactions with atoms outside your body, but these can also, in principle, be included in the calculation.

What this is suggesting is that the future of the Universe is already decided by the current state of the Universe! See page 121 for more details.

Concepts of heat

PHLOGISTON / CALORIC

What happens when a cold block of metal is put into hot water? Everybody knows that the cold block will warm up and the hot water will cool down. Most people would agree that *heat flows* from the hot water into the block. But what is heat? At the beginning of the eighteenth century, a popular theory was that heat was an invisible fluid called **caloric** (or **phlogiston**). The more caloric an object contained the hotter it was. Although scientists do not currently accept this theory, often people still use language that implies heat is something physical that flows from hot to cold.

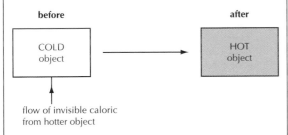

In many situations caloric theory can be used to explain the observations made – particularly when a hot and a cold object are involved. One situation that causes a little difficulty is warming that takes place as a result of friction. If things are warmed because of caloric, where has this heat come from? An accepted explanation was that caloric was generated as a result of friction. It could have been released from the materials being rubbed.

RUMFORD'S EXPERIMENTS

Benjamin Thompson (1753 – 1814), more usually known as Count Rumford, undertook a series of investigations into the process by which friction could produce heating. Although an American by birth, he ended up being employed by the reigning family in Bavaria where he gained the title of Count Rumford. His essay, "An Inquiry Concerning the Source of the Heat Which Is Excited by Friction" was read before the Royal Society on January 25, 1798.

When a metal cannon is manufactured, a hole is bored in a solid piece of metal. He noticed that the metal got very hot. Where was the heat coming from? One possibility was that caloric was being released from the metal during the cutting process. If this was the case then the heat capacity of these chips should be different to the original material. A simple comparison did not find any difference.

In a set of careful experiments, he measured the temperature change as a result of using a blunt borer. If the caloric came from the metal then a cutter that only makes a small hole should release less caloric than a sharp borer that makes a big hole. His experiment surrounded the borer and the end of the cannon with a water-filled box.

He managed to show that a blunt cutter produced a very large heating effect. He collected the tiny amount of metallic dust that had been created and was able to conclude that this could not have been the source of the caloric. In further experiments he managed to rule out the air, the water and the rest of the machinery as a possible source. He also made the key observation that the heat generated by friction seemed to be inexhaustible. If this was the case, then the idea that heat was a fluid seemed to be impossible. He concluded that heat should be considered as a form of motion.

JOULE'S EXPERIMENT

James Prescott Joule (1818 – 1889) demonstrated the link between mechanical work and temperature change. He undertook a series of experiments in which he showed that the result of doing mechanical work on a substance could be an increase in temperature. Most importantly he showed that the temperature rise achieved was proportional to the work that had been done once, of course, experimental uncertainties had been taken into consideration.

In many of his experiments, the work was done by a falling weight. This was used variously to force water to flow through narrow tubes, to compress air in a cylinder and to stir up water in a cylinder. In each case a temperature change was produced. It was by careful measurement of this last arrangement that Joule managed to publish what he called the **mechanical equivalent of heat**. In other words he was able to find the amount of work that needed to be done in order to raise the temperature of water by a fixed amount.

In this experiment, turning the handle raises the weights. As they fall, the paddles rotate and churn up the water. A small temperature rise can be recorded for the water. By careful repetition he was able to show that a given temperature change required a specific amount of work to be done. His measurements were very precise. He even attempted to take into account the sound generated when the weight hit the ground.

Joule was confident that the experimental work was sufficiently accurate to be able to finish his experimental descriptions with a statement of the amount of work needed (using the units that were used at the time). He stated that "the quantity of heat capable of increasing the temperature of a pound of water by 1° Fahrenheit requires for its evolution the expenditure of a mechanical force represented by the fall of 772 pounds through the space of one foot".

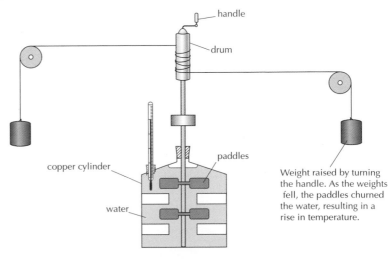

Joule's mechanical equivalent experiment

Discovery of natural electrification and magnetism

ELECTRICAL EFFECTS

The word 'electricity' comes from the Greek word for the material amber. Amber is a naturally occurring material that is golden-coloured. It has long been used to make jewellery. It is, in fact, the fossilised sap from pine trees and occasionally pieces have been found with fossilised insects visible inside them.

Owners of amber jewellery know that it tends to get dusty very easily. Thales of Miletos (625–547 BC) had observed that, when rubbed, amber attracted light objects such as dust. This process of electrification by friction was not successfully experimentally investigated until much later. Indeed it was not until William Gilbert (1544–1603) published his book *De Magnete Magneticisque Corporibus et de Magno Magnete Tellure Physiologia Nova* (1600) that the distinction between electricity and magnetism was established.

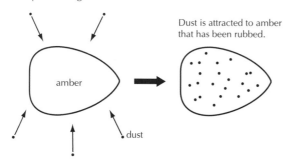

Dust is attracted to amber that has been rubbed.

Electrification of amber

Around 1660, Otto von Guerick (1602–1686) invented an apparatus that opened the way for a great deal of subsequent research – the first electrostatic generator. This consisted of a globe of sulphur that rotated on an axis and was electrified by friction. Using this charged globe he was able to demonstrate the attraction between the charged globe and neutral objects such as paper, leaves or foil. He was also able to observe electrostatic repulsion. He saw that objects that were initially attracted to the globe were then repelled away – this repulsion continued until they touched some other body.

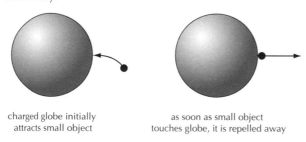

charged globe initially attracts small object

as soon as small object touches globe, it is repelled away

Attraction and repulsion

Stephen Gray (1666–1736) was the first to develop the idea of insulators and conductors from his experimental work with electrostatic effects. He also discovered the phenomenon of remote influence and the electrostatic induction effect.

MAGNETIC EFFECTS

The first known mention of lodestone – a naturally occurring magnet, iron oxide – comes from China. In the first century BC, Chinese fortune-tellers began using lodestone to construct their "divining" boards. Within a hundred years this had led to the creation of the first compasses. The Chinese also discovered that iron was a magnetic material and that lodestones could be used to turn small iron needles into compass magnets.

In the West, analysis of magnets and compasses took place at a much later date. The first Western reference to compasses used for navigation was around about the twelfth century in Alexander Neckam's *De naturis rerum*. Although the use of compasses, and thus the magnet effect, had been around for some time, it was William Gilbert's experiments in the sixteenth century with lodestones, magnets and electrical materials that first introduced the idea of the Earth behaving like a great magnet.

DU FAY

Charles Francois de Cisternay du Fay (1698 – 1739) was the first to clearly state the 'two-fluid' theory of electricity. In essence, he identified what he called two different electricities. He called one type 'vitreous electricity'. This was the type of electricity that existed on a glass rod when it was rubbed with silk. The other type was 'resinous electricity'. This existed on a rubber rod when rubbed with wool. He observed that when electricity was passed between two objects by contact, the electricity would remain of the same original type.

He also recognised that objects with different electricities attracted whereas those with similar electricities repelled. From the modern perspective, he had discovered that electric charges are of two types and that like charges repel while unlike charges attract. He had, however, failed to identify any link between the two types of electricity.

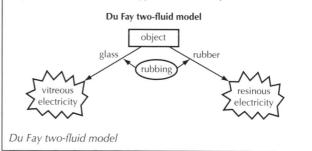

Du Fay two-fluid model

The concepts of electric charge, electric force and electric field

FRANKLIN

Benjamin Franklin (1706 – 1790) is popularly known for flying a kite during a thunderstorm. As a result of the experiment he was able to show that lightning is a similar kind of electricity to that involved in electrostatic experiments.

He also proposed the 'one-fluid' theory of electricity, to which we owe the names positive and negative.

He suggested that if one object is made plus by friction then the other object involved must be minus. In other words friction transfers electric fluid from one object to another.

In modern terms, friction causes the negative electrons to be transfered from one object to another.

Franklin's one-fluid model does present problems, however. Before the fluid flowed between the objects, they were both neutral. If charging is just the transfer of a fluid, then this would suggest that the repulsion is present within an uncharged object. His terminology of positive and negative has remained even if his single-fluid model has now been rejected.

Franklin was also involved with furthering the understanding of the mathematical nature of the electrostatic force.

PRIESTLEY

Joseph Priestley (1733–1804) was able to deduce the inverse square law for electric charges. This law states that the force between two charges is proportional to the inverse of the square of the distances between the charges.

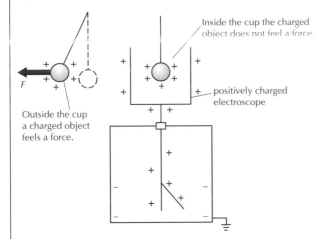

He analysed an experiment that showed the absence of electrical effects inside a charged, hollow, conducting object.

Priestley considered the case of a small charge P placed inside a hollow, conducting sphere.

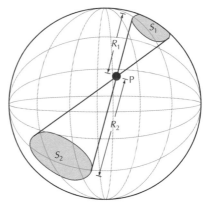

- the ratio of the two areas is given by $\frac{S_1}{S_2} = \frac{R_1{}^2}{R_2{}^2}$
- if the charge is uniformly distributed, the amount of charge is proportional to the area.
- if the net force on P is zero, then the force due to S_1 is equal and opposite to S_2.
- this means that the forces must be proportional to the charge and inversely proportional to the (distance)^2.
- all parts of the surface can be paired off in this way so the net force on P will always be zero.

It should be noted that his approach is an indirect one. Although the mathematical relationship can be deduced from the observations, Priestley did not actually experimentally measure the force. Coulomb confirmed his deduction using a direct approach.

COULOMB

Charles Augustin de Coulomb (1736–1806) developed an experimental arrangement by which he could directly confirm the inverse square law of electric charges. In essence it consisted of a torsion balance. Coulomb twisted the torsion head to maintain equilibrium. He had earlier determined a relationship between the angle of twist and the force that must be applied.

He also found that the force was proportional to the product of the charges

$$F \propto Q_1 Q_2$$

and was inversely proportional to the (separation of the charges)^2 $\qquad F \propto \frac{1}{R^2}$

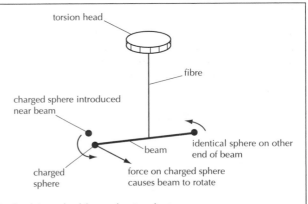

Coulomb's method for evaluating the inverse square

Magnetic effects of electric currents and electric effects of magnetic fields

OERSTED

Hans Christian Oersted (1777–1851) was the first to note that an electric current caused the deflection of a magnetic compass needle. He discovered the effect by accident while performing a physics demonstration. He had planned to show the heating of a wire by an electric current and also to carry out demonstrations of magnetism, for which he had prepared a large compass needle mounted on a wooden stand.

While performing his electric demonstration, Oersted observed that every time the electric current was switched on, the compass needle moved. He finished the demonstrations without making comment, but in the months that followed worked hard trying to make sense out of the new phenomenon.

It should be noted that the needle was neither attracted to the wire nor repelled from it. Instead, it tended to turn at right angles

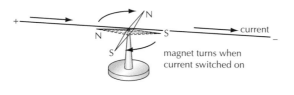

Oersted's experiment

AMPÈRE

André Marie Ampère (1775–1836) independently verified Oersted's observations, but he went on to study the phenomena in more detail. If a current-carrying wire was able to exert a magnetic force, then he supposed that two such wires should be able to interact magnetically.

In a series of ingenious experiments he showed that this interaction existed. He found that two parallel currents attract, whereas two antiparallel currents repel. On top of this, the magnitude of the force between the two long straight parallel currents was mathematically predicable. It was inversely proportional to the distance between the wires (not the distance [2]) and proportional to the value of the current flowing in each.

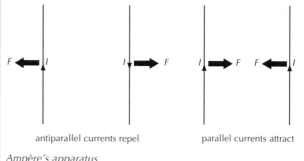

antiparallel currents repel   parallel currents attract

Ampère's apparatus

FARADAY

Michael Faraday (1791–1867) was able to build on the work of Oersted and Ampère and establish a relationship between electric and magnetic effects. His first discovery was to show how the electric and magnetic forces could result in mechanical motion. In principle he was able to build the first electric motor. He also demonstrated that the effects were related. If a current flowed in a wire close to a magnet, which was not fixed down, the magnet would move. If the wire was free to move, but the magnet was fixed, then the wire would move.

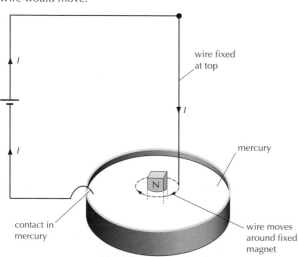

Faraday's electromagnetic motion experiment

He was convinced that it should be possible to create an electric current using magnets – electromagnetic induction. Using a ring of soft iron wrapped in two windings of insulated wire he connected one wire to a sensitive ammeter (a galvanometer). The other wire was connected to a battery.

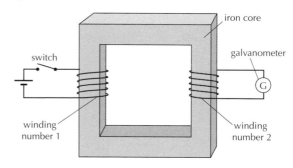

Faraday's electromagnetic induction experiment

When the circuit was broken and connected the galvanometer recorded the pulse of an electric current. He realised that an induced current was produced when the intensity of the magnetic field was changing. This is the basis of a dynamo.

The electron

CROOKES

Sir William Crookes (1832–1919) paved the way for the discovery of the electron. In the 1870s he was investigating the flow of electricity through very low-pressure gases. When conduction occurs in gases, regions of the conducting gas are observed to emit light. The colours and the patterns in the conducting gas depend on the gas used.

On top of the observation of the light in the gas, the glass behind the anode is seen to glow as well. This suggests that something is travelling from the cathode to the anode that is able to provide the gas with energy. This was assumed to be some unknown type of radiation called **cathode rays**. On balance, Crookes felt that the light given off by the gas was a result of collisions of particles or molecules with the surface of the glass.

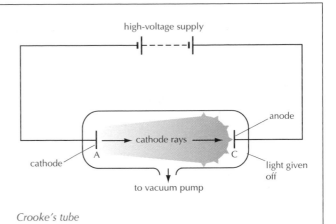

Crooke's tube

HERTZ

Hertz (1857–1894) felt that that these cathode rays had a wave nature and were thus possibly similar to ultraviolet light. Although Crookes had shown that the rays could be deviated by a magnetic field, Hertz noted that the rays were undeviated when they passed between two charged plates. This apparent lack of deviation by an electric field suggested that the cathode rays were waves.

LENARD

Philipp Eduard Anton von Lenard (1862–1947) did an experiment to find out whether cathode rays would, like ultraviolet light, pass through a quartz window in the wall of a discharge tube. He found that they would not do this.

He was eventually able to develop a cathode-ray tube with a thin aluminium window that was thick enough to maintain the vacuum inside, but thin enough to permit the cathode rays to escape.

Lenard observed that cathode rays continued on through the air for about 10 cm or so, and they travelled in a vacuum for several metres without being weakened. He was able to show that if they were particles and did have mass, it was much less than the mass of an atom.

PERRIN

Jean Perrin (1870–1942) showed that the cathode rays appeared to have a negative charge. If the ray was deflected to hit a metal plate, this could cause the leaf on an electroscope to go up as a result of the negative charge collected.

This experiment suggested that the cathode rays were negative particles, but if this was the case they should have been deflected by an electric field. Hertz had shown that they were undeflected by such a field.

THOMSON

Professor Joseph John Thomson (1856–1940) realised that a beam of particles would not be deflected by an electric field if they were surrounded by a conductor. He suspected that the traces of gas remaining in the cathode-ray tube were being turned into an electrical conductor by the cathode rays themselves. To test this idea, he undertook to remove as much of the gas as he could from a tube. He found that in this tube, the cathode rays did bend in an electric field. He was now convinced that they were particles with negative charge.

In an ingenious experiment, he was able to measure the charge-to-mass ratio $\left(\frac{e}{m}\right)$ for these particles (which he called 'corpuscles').

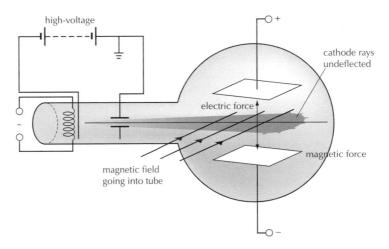

The idea is to balance the electric and magnetic deflections of the cathode rays so that there is no overall deflection. At this point,

magnetic force on electron = electric force on the electron

$$Bev = Ee$$

Thomson calculated the velocity of electrons from the deflection caused by the electric field alone.

The experiment was repeated and the ratio turned out to be independent of the gas used in the tube. The ratio was over a thousand times smaller than the charge-to-mass ratio for a charged hydrogen atom. He proposed that the corpuscles were smaller particles that made up the atom.

The atom and the nucleus

THOMSON'S PLUM PUDDING

Thomson and others had suggested that all atoms contained electrons. These were much smaller than an atom and had been shown to be negative. Since atoms are known to be neutral, any model of an atom had to include an equal amount of positive material to balance the negative charge on the electrons.

Thomson proposed a model in which the tiny electrons were held together in an atom by a positive 'glue'. This is sometimes known as the 'plum pudding' model. The electrons are held in their place in the atom in the same way that the plums are held in place by the dough.

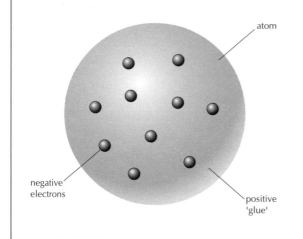

CHADWICK

Even though Rutherford (1871–1937) had identified a positive particle that seemed to be part of the nucleus, it was known that this could not be the only particle in the nucleus. At the very least, a further neutral particle was necessary to explain the observed masses of atoms. It was James Chadwick (1847–1922), a colleague of Rutherford's, who first experimentally determined the mass of the neutron. He was awarded the Nobel prize for his work.

Walter Bothe and Herbert Becker had already seen that when alpha particles bombarded beryllium, an uncharged radiation was emitted. Irene Curie-Joliot and Frederic Joliot found that if this radiation hit a material that contained hydrogen, such as paraffin wax, it caused the emission of protons. The uncharged radiation was assumed to be gamma radiation, even though gamma radiation had not previously been observed to cause the emission of protons.

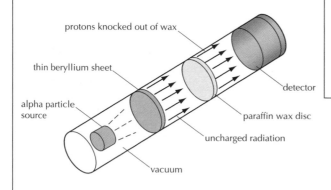

RUTHERFORD'S NUCLEAR MODEL

Rutherford's nuclear model of the atom was developed as a result of the surprising results achieved by two of his research students, Hans Geiger and Ernst Marsden, when investigating the scattering of alpha particles – see page 48 for more details. It should be noted that the model is more than just a vague description of a possible structure for atoms. It allowed Rutherford to develop a mathematical expression for the relative numbers of alpha particles that should be deflected through any given angle. Geiger and Marsden showed that these predictions were experimentally verified.

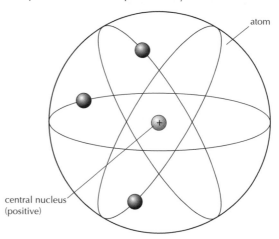

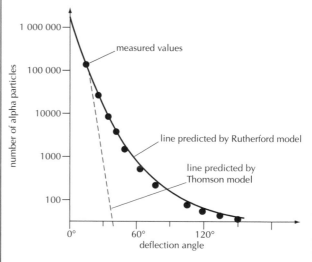

Rutherford also went on to discover one of the fundamental building blocks of the nucleus – the proton. Although the details of this experiment are not needed for the IB Physics syllabus, it is interesting to note that its discovery also involved alpha particles. If an alpha particle has the right energy, a nitrogen nucleus can absorb the alpha particle which causes the emission of a proton.

$$^{14}_{7}N + {}^{4}_{2}He \rightarrow {}^{17}_{8}O + {}^{1}_{1}p$$

Using the laws of conservation of energy and momentum, Chadwick was able to show that the uncharged radiation consisted of particles of mass 1 and charge 0 – neutrons. In other words the original reaction must have been as follows.

$$^{9}_{4}Be + {}^{4}_{2}He \rightarrow {}^{12}_{6}C + {}^{1}_{0}n$$

Atomic spectra and the Bohr model of the atom

HYDROGEN SPECTRUM

The emission spectrum of atomic hydrogen consists of particular wavelengths. In 1885 a Swiss schoolteacher called Johann Jacob Balmer found that the visible wavelengths fitted a mathematical formula.

These wavelengths, known as the **Balmer series**, were later shown to be just one of several similar series of possible wavelengths that all had similar formulae. These can be expressed in one overall formula called the **Rydberg formula**.

$$\frac{1}{\lambda} = R_H \left(\frac{1}{n^2} - \frac{1}{m^2} \right)$$

λ – the wavelength

m – a whole number larger than 2 i.e. 3, 4, 5 etc

For the **Lyman series** of lines (in the ultraviolet range) $n = 1$. For the **Balmer series** $n = 2$. The other series are the **Paschen** ($n = 3$), **Brackett** ($n = 4$), and the **Pfund** ($n = 5$) series. In each case the constant R_H, called the **Rydberg constant**, has the one unique value, $1.097 \times 10^7 \, \text{m}^{-1}$.

BOHR MODEL

Niels Bohr (1885 – 1962) developed a model of hydrogen that is outlined on page 83.

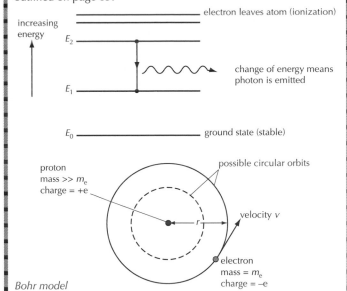

Bohr model

Bohr postulated that:

- An electron does not radiate energy when in a stable orbit. The only stable orbits possible for the electron are ones where the **angular momentum** of the orbit is an integral multiple of $\frac{h}{2\pi}$ where h is a fixed number (6.6×10^{-34} J s) called **Planck's constant**. Mathematically

$$m_e v r = \frac{nh}{2\pi}$$

[angular momentum is equal to $m_e v r$]

- When electrons move between stable orbits they radiate (or absorb) energy.

$$F_{\text{electrostatic}} = \text{centripetal force}$$

$$\therefore \quad \frac{e^2}{4\pi\varepsilon_0 r^2} = \frac{m_e v^2}{r}$$

but $\quad v = \frac{nh}{2\pi \, m_e r}$ [from 1st postulate]

$$\therefore \quad r = \frac{\varepsilon_0 n^2 h^2}{\pi m_e e^2}$$

Total energy of electron $= KE + PE$

where $\quad KE = \frac{1}{2} m_e v^2 = \frac{1}{2} \frac{e^2}{(4\pi\varepsilon_0 r)}$

and $\quad PE = -\frac{e^2}{4\pi\varepsilon_0 r}$ [electrostatic PE]

so total energy $\quad E_n = -\frac{1}{2} \frac{e^2}{4\pi\varepsilon_0 r} = -\frac{m_e e^4}{8\varepsilon_0 n^2 h^2}$

This final equation shows that:

- the electron is bound to (= 'trapped by') the proton because overall it has negative energy.
- the energy of an orbit is proportional to $-\frac{1}{n^2}$. In electronvolts

$$E_n = -\frac{13.6}{n^2}$$

The second postulate can be used (with the full equation) to predict the wavelength of radiation emitted when an electron makes a transition between stable orbits.

$$hf = E_2 - E_1$$

$$= \frac{m_e e^4}{8\varepsilon_0^2 h^2} \left(\frac{1}{n_1^2} - \frac{1}{n_2^2} \right)$$

but $\quad f = \frac{c}{\lambda}$

$$\therefore \quad \frac{1}{\lambda} = \frac{m_e e^4}{8\varepsilon_0^2 c h^3} \left(\frac{1}{n_1^2} - \frac{1}{n_2^2} \right)$$

It should be noted that:
- this equation is of the same form as the Rydberg formula.
- the values predicted by this equation are in very good agreement with experimental measurement.
- the Rydberg constant can be calculated from other (known) constants. Again the agreement with experimental data is good.

The limitations to this model are:
- if the same approach is used to predict the emission spectra of other elements, it fails to predict the correct values for atoms or ions with more than one electron.
- the first postulate (about angular momentum) has no theoretical justification.
- theory predicts that electrons should, in fact, not be stable in circular orbits around a nucleus. Any accelerated electron should radiate energy. An electron in a circular orbit is accelerating so it should radiate energy and thus spiral in to the nucleus.
- it is unable to account for relative intensity of the different lines.
- it is unable to account for the fine structure of the spectral lines.

The de Broglie hypothesis and the Schrödinger model of the atom

dE BROGLIE AND MATTER WAVES

Louis Victor Pierre Raymond duc de Broglie (1892–1987) developed the idea of matter waves as a result of studies into the particle nature of waves. He then extended this wave – particle duality to material particles and, in particular, electrons.

If wave and particle properties were to apply to electrons, then there should be a wavelength associated with an electron in the same way that there is a momentum associated with a photon of light energy.

photon energy $E = hf = mc^2$

photon momentum $p = mc$

This means that $p = \dfrac{hf}{c} = \dfrac{h}{\lambda}$

In other words, $\lambda = \dfrac{h}{p}$

SCHRÖDINGER MODEL

Erwin Schrödinger (1887–1961) built on the concept of matter waves and proposed an alternative model of the hydrogen atom using wave mechanics. The **Copenhagen interpretation** is a way to give a physical meaning to the mathematics of wave mechanics.

- The description of particles (matter and /or radiation) in quantum mechanics is in terms of a **wave function** ψ. This wave function has no physical meaning but the square of the wave function does.
- At any instant of time, the wave function has different values at different points in space.
- The mathematics of how this wave function develops with time and interacts with other wave functions is like the mathematics of a travelling wave.
- The probability of finding the particle (electron or photon etc.) at any point in space within the atom is given by the square of the amplitude of the wave function at that point.
- When an observation is made the wave function is said to **collapse**, and the complete physical particle (electron or photon etc.) will be observed to be at one location.

In Schrödinger's model there are different wave functions depending on the total energy of the electron. Only a few particular energies result in wave functions that fit the boundary conditions – electrons can only have these particular energies within an atom. An electron in the ground state has a total energy of -13.6 eV, but its position at any given time is undefined in this model. The wave function for an electron of this energy can be used to calculate the probability of finding it at a given distance away from the nucleus. This is achieved as follows:

- The wavelength of the wave function varies with the momentum of the electron as given by the de Broglie relationship. $\lambda = \dfrac{h}{p}$
- The total energy of an electron is known to be the sum of its kinetic energy and its potential energy – these both vary with distance.

- As the distance away from the nucleus increases, an electron's kinetic energy and thus its momentum would also decrease. This means the wavelength will increase with distance away from the nucleus.
- The resulting **orbital** for the electron can be described in terms of the probability of finding the electron at a certain distance away. The probability of finding the electron at a given distance away is shown in the graph below.

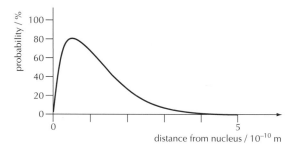

- The electron in this orbital can be visualised as a 'cloud' of varying electron density. It is more likely to be in some places than other places, but its actual position in space is undefined.

There are other fixed total energies for the electron that result in different possible orbitals. In general as the energy of the electron is increased it is more likely to be found at a further distance away from the nucleus.

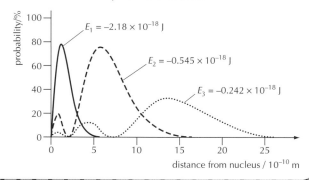

COMPARISONS OF MODELS OF THE HYDROGEN ATOM

	Bohr model	Schrödinger's model
Assumptions	No theoretical justification is provided for 1. the rule concerning angular momentum 2. lack of radiation from the accelerating electron	Wave – particle duality is assumed to apply to all particles.
Applicability	Easy to apply	Mathematics very difficult to solve.
Limitations	Unable to predict spectra for atoms or ions with more than one electron.	Model works, in principle, for all atoms.
Successes	Predicts some aspects of atomic H spectrum.	Predicts all aspects of atomic H spectrum accurately.

The Heisenberg uncertainty principle and the loss of determinism

HEISENBERG

Werner Heisenberg (1901–1976) was one of the leading developers of quantum mechanics. Before Schrödinger's model had been proposed, he had already co-authored a paper that introduced a new mathematical approach based on the use of matrices ('matrix mechanics'). This version of quantum mechanics accounted for many of the properties of atoms but lacked the physical interpretation provided by Schrödinger's model.

The **Heisenberg uncertainty principle** identifies a fundamental limit to the possible accuracy of any physical measurement. This limit arises because of the nature of quantum mechanics and not as a result of the ability (or otherwise) of any given experimenter. He showed that it was impossible to measure exactly the position **and** the momentum of a particle simultaneously. The more precisely the position is determined, the less precisely the momentum is known in this instant, and vice versa. They are linked variables.

There is a mathematical relationship linking these uncertainties.

$$\Delta x \Delta p \geq \frac{h}{4\pi}$$

Δx The uncertainty in the measurement of position

Δp The uncertainty in the measurement of momentum

Measurements of energy and time are also linked variables.

$$\Delta E \Delta t \geq \frac{h}{4\pi}$$

ΔE The uncertainty in the measurement of energy

Δt The uncertainty in the measurement of time

The implications of this lack of precision are profound. On page 112 it was shown that a deterministic theory allows us (in principle) to make absolute predictions about the future. Before quantum theory was introduced, the physical world was best described by deterministic theories – e.g. Newton's laws.

Quantum mechanics is not deterministic. It cannot ever predict exactly the results of a single experiment. It only gives us the probabilities of the various possible outcomes. The uncertainty principle takes this even further. Since we cannot know the precise position and momentum of a particle at any given time, its future can never be determined precisely. The best we can do is to work out a range of possibilities for its future.

It has been suggested that science would allow us to calculate the future so long as we know the present exactly. As Heisenberg himself said, it is not the conclusion that is wrong but the premise.

IB QUESTIONS – OPTION E – THE HISTORY AND DEVELOPMENT OF PHYSICS

1 Count Rumford observed the barrels of iron cannons being bored out by cutting tools, producing metal chips. He reported in 1798 that this mechanical process seemed to provide an 'inexhaustible supply of heat'.

(a) How did the 'caloric' theory existing at that time account for the production of heat in the cutting process? [2]

(b) If the boring tool became blunt so that it did not cut as well, what did the caloric theory predict would happen to the rate of heat production, and why? Was this observed? [2]

(c) Why was the observation that the supply of heat seemed 'inexhaustible' a problem for the caloric theory? [2]

(d) What new idea did Rumford propose to account for the production of heat in this process? [2]

2 This question is about the quantum concept.

A biography of Schrödinger contains the following sentence: "Shortly after de Broglie introduced the concept of *matter waves* in 1924, Schrödinger began to develop a new atomic theory."

(a) Explain the term '*matter waves*'. State what quantity determines the wavelength of such waves. [2]

(b) Electron diffraction provides evidence to support the existence of matter waves. What is electron diffraction? [2]

(c) Calculate the de Broglie wavelength of electrons with a kinetic energy 30 eV. [3]

(d) How does the concept of *matter waves* apply to the electrons within an atom? [2]

3 This question is about Newton's contribution to the understanding of gravitation.

(a) In order to solve many problems in mechanics, Newton invented the branch of mathematics we now call *calculus*. However, in the *Principia* he did not use this new mathematics. What was the predominant form of mathematics that he used? [1]

(b) One of Newton's important contributions in the field of mechanics was to realise that the force of gravity was *universal*. What is meant by the phrase 'the force of gravity is universal'? [2]

(c) Newton concluded that the same force was responsible for the acceleration of a falling object on the Earth's surface and for the acceleration **of the Moon** in orbit. This he showed would be true as long as that force was a 'central force' and followed an 'inverse square law'. In this context, explain what is meant by a 'central force' and by 'an inverse square law'. [2]

(d) The following data was well known in Newton's time (although not so precisely):

acceleration due to gravity
at the Earth's surface $= 9.81$ m s^{-2}
Earth's radius $= 6.37 \times 10^6$ m
Earth – Moon distance $= 3.84 \times 10^8$ m
Moon's period $= 27.3$ days

We can repeat a part of Newton's work using the above data, and **only this data**.

(i) Assuming the force of gravity is universal and obeys an inverse square law, calculate the acceleration expected at the Moon's orbit, due to this force. [2]

(ii) Calculate the centripetal acceleration required for the Moon to maintain its orbit. [3]

(iii) In order for Newton to arrive at the conclusion he did, what must be true about the two accelerations calculated in **(i)** and **(ii)** above? [1]

4 This question is about models of the atom.

Two models of the atom are based on Bohr's and Schrödinger's theories. Such models are devised to explain observed atomic phenomena.

(a) In what ways do the Bohr and Schrödinger models of the hydrogen atom differ? [3]

(b) How does the existence of discrete atomic energy levels arise in each model? Give explanations in physical terms rather than mathematical.

Explanation from Bohr model [2]

Explanation from Schrödinger model [2]

(c) Both models can at least explain the observed spectra of hydrogen. Why then is one model preferred to the other? In what respects is the model arising from Schrödinger theory deemed superior to Bohr's model? [3]

The Solar System and beyond

SOLAR SYSTEM

We live on the Earth. This is one of nine planets that orbit the Sun – collectively this system is known as the Solar System. Each planet is kept in its elliptical orbit by the gravitational attraction between the Sun and the planet.

	Mercury	Venus	Earth	Mars	Jupiter	Saturn	Uranus	Neptune	Pluto
diameter / km	4880	12104	12756	6787	142800	120000	51800	49500	2300
distance to Sun / $\times 10^8$ m	58	107.5	149.6	228	778	1427	2870	4497	5900

Relative positions of the planets

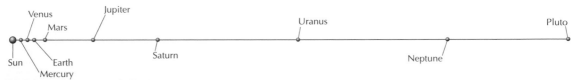

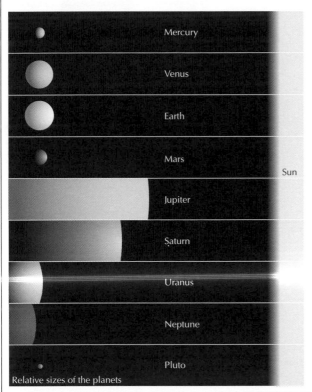

Relative sizes of the planets

Some of these planets (including the Earth) have other small planets orbiting around them called moons. Our Moon is about 1/400th of the size of the Earth and 3.8×10^8 m away.

An **asteroid** is a small rocky body that drifts around the Solar System. There are many orbiting the sun between Mars and Jupiter – the asteroid belt. An asteroid on a collision course with another planet is known as a meteoroid.

Small meteors can be vaporised due to the friction with the atmosphere ('shooting stars') whereas larger ones can land on Earth. The bits that arrive are called **meteorites**.

Comets are mixtures of rock and ice (a 'dirty snowball') in very elliptical orbits around the Sun. Their 'tails' always point away from the Sun.

THE UNIVERSE

Our Sun is just one of the billions of stars in our **galaxy** (the Milky Way galaxy). The galaxy rotates with a period of about 2×10^8 years.

Beyond our galaxy, there are billions of other galaxies. Some of them are grouped together into **clusters** or **super clusters** of galaxies, but the vast majority of space (like the gaps between the planets or between stars) appears to be empty – essentially a vacuum. Everything together is known as the **Universe**.

	1.5×10^{26} m (= 15 billion light years)	the visible Universe
	5×10^{22} m (= 5 million light years)	local group of galaxies
	10^{21} m (= 100 000 light years)	our galaxy
	10^{13} m (= 0.001 light years)	our Solar System

UNITS

When comparing distances on the astronomical scale, it can be quite unhelpful to remain in SI units. Possible other units include the **astromonical unit (AU)**, the **parsec (pc)** or the **light year (ly)**. See page 129 for the definition of the first two of these.

The light year is the distance travelled by light in one year (9.5×10^{15} m). The nearest star to our Sun is about 4 light years away. Our galaxy is about 100 000 light years across. The nearest galaxy is about a million light years away and the observable Universe is about 1.5×10^9 light years across.

VIEW FROM EARTH

If we look up at the night sky we see the stars – many of these 'stars' are, in fact, other galaxies but they are very far away. The stars in our own galaxy appear as a band across the sky – the Milky Way.

Patterns of stars have been identified and 88 different regions of the sky have been labelled as the different constellations.

Over the period of a night, the constellations seem to rotate around one star. This apparent rotation is a result of the rotation of the Earth about its own axis.

On top of this nightly rotation, there is a slow change in the stars and constellations that are visible from one night to the next. This variation over the period of one year is due to the rotation of the Earth about the Sun.

Energy source

ENERGY FLOW FOR STARS

The stars are emitting a great deal of energy. The source for all this energy is the fusion of hydrogen into helium. See page 54. Sometimes this is referred to as 'hydrogen burning' but it this is not a precise term. The reaction is a nuclear reaction, not a chemical one (such as combustion). Overall the reaction is

$$4\,^1_1p \rightarrow \,^4_2He + 2\,^0_1e^+ + 2\nu$$

The mass of the products is less than the mass of the reactants. Using $E = m\,c^2$ we can work out that the Sun is losing mass at a rate of 4×10^9 kg s^{-1}. This takes place in the core of a star. Eventually all this energy is radiated from the surface – approximately 10^{26} J every second. The structure inside a star does not need to be known in detail.

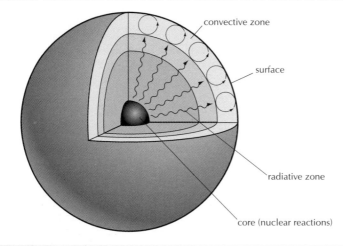

- convective zone
- surface
- radiative zone
- core (nuclear reactions)

EQUILIBRIUM

The Sun has been radiating energy for the past 4 ½ billion years. It might be imagined that the powerful reactions in the core should have forced away the outer layers of the Sun a long time ago. Like other stars, the Sun is stable because there is an equilibrium between this outward pressure and the inward gravitational force.

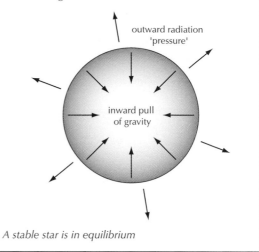

outward radiation 'pressure'

inward pull of gravity

A stable star is in equilibrium

Luminosity

LUMINOSITY AND APPARENT BRIGHTNESS

The total power **radiated** by a star is called its **luminosity**. The SI units are watts. This is very different to the power **received** by an observer on the Earth. The power received per unit area is called the **brightness** of the star. The SI units are W m^{-2}.

If two stars were at the **same distance** away from the Earth then the one with the greatest luminosity would be brighter. Stars are, however, at different distances from the Earth. The brightness is inversely proportional to the (distance)^2.

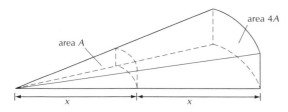

area A

area $4A$

x x

As distance increases, the brightness decreases since the light is spread over a bigger area.

distance	brightness
x	b
$2x$	$\frac{b}{4}$
$3x$	$\frac{b}{9}$
$4x$	$\frac{b}{16}$
$5x$	$\frac{b}{25}$
and so on	

inverse square

Brightness $b = \dfrac{L}{4\pi r^2}$

It is thus possible for two very different stars to have the same apparent brightness. It all depends on how far away the stars are.

close star (small luminosity)

distant star (high luminosity)

Two stars can have the same apparent brightness even if they have different luminosities

UNITS

The SI units for luminosity and brightness have already been introduced. In practice astronomers compare the brightness of stars using the **apparent magnitude** scale. This scale is defined on page 130. A magnitude 1 star is brighter than a magnitude 3 star. This brightness is often shown on star maps.

The magnitude scale can also be used to compare the luminosity of different stars, provided the distance to the star is taken into account. Astronomers quote values of **absolute magnitude** in order to compare luminosities on an easy and familiar scale.

Wien's law and the Stefan – Boltzmann law

BLACK-BODY RADIATION

In general, the radiation given out from a hot object depends on many things. It is possible to come up with a theoretical model for the 'perfect' emitter of radiation. The 'perfect' emitter will also be a perfect absorber of radiation – a black object absorbs all of the light energy falling on it. For this reason the radiation from a theoretical 'perfect' emitter is known as **black-body radiation**.

Black-body radiation does not depend on the nature of the emitting surface, but it does depend upon its temperature. At any given temperature there will be a range of different wavelengths (and hence frequencies) of radiation that are emitted. Some wavelengths will be more intense than others. This variation is shown in the graphs below.

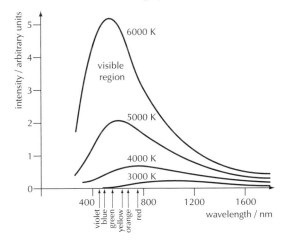

To be absolutely precise, it is not correct to label the y-axis on the above graph as the intensity, but this is often done. It is actually something that could be called the intensity function. This is defined so that the area under the graph (between two wavelengths) gives the intensity emitted in that wavelength range. The total area under the graph is thus a measure of the total power radiated.

Although stars are not perfect emitters, their radiation spectrum is approximately the same as black-body radiation.

WIEN'S LAW

Wien's displacement law relates the wavelength at which the intensity of the radiation is a maximum λ_{max} to the temperature of the black body T. This states that

$\lambda_{max} T = \text{constant}$

The value of the constant can be found by experiment. It is 2.9×10^{-3} m K. It should be noted that in order to use this constant, the wavelength should be substituted into the equation in metres and the temperature in kelvin.

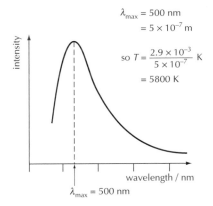

The peak wavelength from the Sun is approximately 500 nm.

$\lambda_{max} = 500 \text{ nm}$
$= 5 \times 10^{-7} \text{ m}$

so $T = \dfrac{2.9 \times 10^{-3}}{5 \times 10^{-7}}$ K

$= 5800 \text{ K}$

We can analyse light from a star and calculate a value for its surface temperature. This will be much less than the temperature in the core. Hot stars will give out all frequencies of visible light and so will tend to appear white in colour. Cooler stars might well only give out the higher wavelengths (lower frequencies) of visible light – they will appear red.

STEFAN – BOLTZMANN LAW

The Stefan – Boltzmann law links the **total** power radiated by a black body (per unit area) to the temperature of the black-body. The important relationship is that

Total power radiated $\propto T^4$

In symbols we have,

Total power radiated $= \sigma A T^4$

Where

σ is a constant called the Stefan – Boltzmann constant. $\sigma = 5.67 \times 10^{-8}$ W m^{-2} K^{-4}

A is the surface area of the emitter (in m^2)

T is the absolute temperature of the emitter (in kelvin)

The radius of the Sun $= 6.96 \times 10^8$ m.

Surface area	$= 4\pi r^2$
	$= 6.09 \times 10^{18}$ m^2
If temperature	$= 5800$ K
then luminosity	$= \sigma A T^4$
	$= 5.67 \times 10^{-8} \times 6.09 \times 10^{18} \times (5800^4)$
	$= 3.9 \times 10^{26}$ W

In terms of stars, A is the surface area of the star. The radius of the star r is linked to its surface area using the equation $A = 4\pi r^2$. This means that if know the luminosity of a star and its temperature, we can work out its size.

Stellar spectra

ABSORPTION LINES

The radiation from stars is not a perfect continuous spectrum – there are particular wavelengths that are 'missing'.

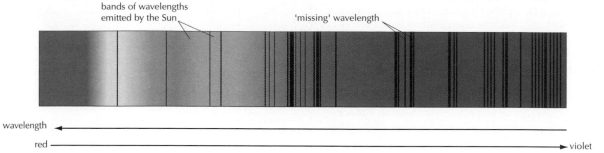

bands of wavelengths
emitted by the Sun

'missing' wavelength

wavelength

red ———————————————————————————————————→ violet

The missing wavelengths correspond to the absorption spectrum of a number of elements. Although it seems sensible to assume that the elements concerned are in the Earth's atmosphere, this assumption is incorrect. The wavelengths would still be absent if light from the star was analysed in space.

The absorption is taking place in the outer layers of the star. This means that we have a way of telling what elements exist in the star – at least in its outer layers.

CLASSIFICATION OF STARS

Different stars give out different spectra of light. This allows us to classify stars by their **spectral class**. Stars that emit the same type of spectrum are allocated to the same spectral class. Historically these were just given a different letter, but we now know that these different letters also correspond to different surface temperatures.

The seven main spectral classes (in order of **decreasing** surface temperature) are O, B, A, F, G, K and M. This is sometimes remembered as 'Oh Be A Fine Girl/Guy, Kiss Me'. The main spectral classes can be subdivided.

Class	Effective surface temperature/K	Colour
O	28 000–50 000	blue
B	9900–28 000	blue-white
A	7400–9900	white
F	6000–7400	yellow-white
G	4900–6000	yellow
K	3500–4900	orange
M	2000–3500	orange-red

SUMMARY

If we know the distance to a star (see page 130 for the techniques) we can analyse the light from the star and work out:

- the chemical composition (by analysing the absorption spectrum).
- the surface temperature (using a measurement of λ_{max} and Wien's law).

If we know the distance we can work out:

- the luminosity (using measurements of the brightness and the distance away).
- the surface area of the star (using the luminosity, the surface temperature and the Stefan – Boltzmann law).

Types of star

SINGLE STARS

The source of energy for our Sun is the fusion of hydrogen into helium. This is also true for many other stars. There are however, other types of stars that are known to exist in the Universe.

Type of star	Description
Red giant stars	As the name suggests, these stars are large in size and red in colour. Since they are red, they are comparatively cool. They turn out to be one of the later possible stages for a star. The source of energy is the fusion of some elements other than hydrogen.
White dwarf stars	As the name suggests, these stars are small in size and white in colour. Since they are white, they are comparatively hot. They turn out to be one of the final stages for some stars. Fusion is no longer taking place, and a white dwarf is just a hot remnant that is cooling down. Eventually it will cease to give out light when it becomes sufficiently cold. It is then known as a brown dwarf.
Cepheid variables	These are stars that are a little unstable. They are observed to have a regular variation in brightness and hence luminosity. This is thought to be due to an oscillation in the size of the star. They are quite rare but are very useful as there is a link between the period of brightness variation and their average luminosity. This means that astronomers can use them to help calculate the distance to some galaxies.

BINARY STARS

Our Sun is a single star. Many 'stars' actually turn out to be two (or more) stars in orbit around each other. (To be precise they orbit around their common centre of mass). These are called **binary stars**.

binary stars – two stars in orbit around their common centres of mass

There are different categories of binary star – **visual**, **spectroscopic** and **eclipsing**.

1. A visual binary is one that can be distinguished as two separate stars using a telescope.

2. A spectroscopic binary star is identified from the analysis of the spectrum of light from the 'star'. Over time the wavelengths show a periodic shift or splitting in frequency. An example of this is shown below.

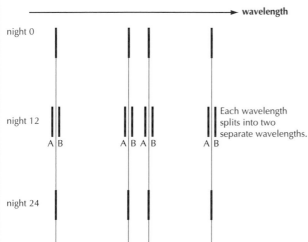

The explanation for the shift in frequencies involves the Doppler effect. As a result of its orbit, the stars are sometimes moving towards the Earth and sometimes they are moving away. When a star is moving towards the Earth, its spectrum will be blueshifted. When it is moving away, it will be redshifted.

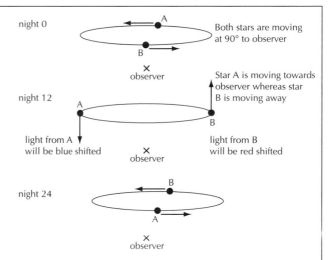

3. An eclipsing binary star is identified from the analysis of the brightness of the light from the 'star'. Over time the brightness shows a periodic variation. An example of this is shown below.

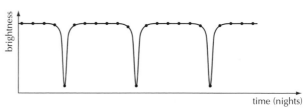

The explanation for the 'dip' in brightness is that as a result of its orbit, one star gets in front of the other. If the stars are of equal brightness, then this would cause the total brightness to drop to 50%.

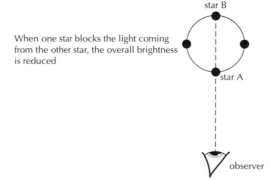

When one star blocks the light coming from the other star, the overall brightness is reduced

The Hertzsprung – Russell diagram

H – R DIAGRAM

The point of classifying the various types of stars is to see if any patterns exist. A useful way of making this comparison is the **Hertzsprung – Russell diagram**. Each dot on the diagram represents a different star. The following axes are used to position the dot.

- The vertical axis is the luminosity (or absolute magnitude) of the star. It should be noted that the scale is not a linear one (it is logarithmic).
- The horizontal axis is the spectral class of the star in the order OBAFGKM. This is the same as a scale of **decreasing** temperature. Once again, the scale is not a linear one.

The result of such a plot is shown below.

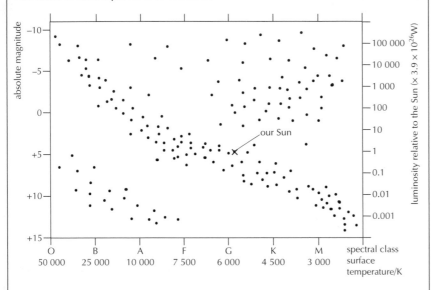

A large number of stars that fall on a line that (roughly) goes from top left to bottom right. This line is known as the **main sequence** and stars that are on it are known as main sequence stars. Our Sun is a main sequence star. These stars are 'normal' stable stars – the only difference between them is their mass. They are fusing hydrogen to helium. The stars that are not on the main sequence can also be broadly put into categories.

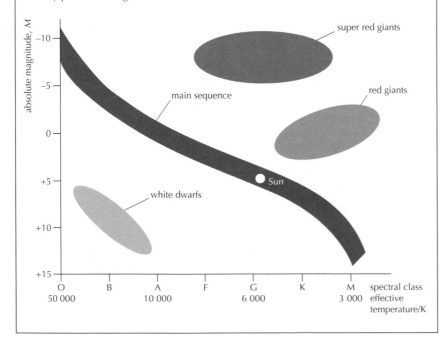

Parallax method

PRINCIPLES OF MEASUREMENT

As you move from one position to another objects change their relative positions. As far as you are concerned, near objects appear to move when compared with far objects. Objects that are very far away do not appear to move at all. You can demonstrate this effect by closing one eye and moving your head from side to side. An object that is near to you (for example the tip of your finger) will appear to move when compared with objects that are far away (for example a distant building).

This apparent movement is known as **parallax** and the effect can used to measure the distance to some of the stars in our galaxy. All stars appear to move over the period of a night, but some stars appear to move in relation to other stars over the period of a year.

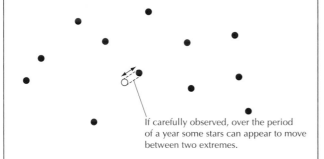

If carefully observed, over the period of a year some stars can appear to move between two extremes.

The reason for this apparent movement is that the Earth has moved over the period of a year. This change in observing position has meant that a close star will have an apparent movement when compared with a more distant set of stars. The closer a star is to the Earth, the greater will be the parallax shift.

Since all stars are very distant, this effect is a very small one and the parallax angle will be very small. It is usual to quote parallax angles not in degrees, but in seconds. An angle of 1 second of arc ('') is equal to one sixtieth of 1 minute of arc (') and 1 minute of arc is equal to one sixtieth of a degree.

In terms of angles, $3600'' = 1°$

$360° = 1$ full circle.

EXAMPLE

The star alpha Eridani (Achemar) is 1.32×10^{18} m away. Calculate its parallax angle.

$d = 1.32 \times 10^{18}$ m

$= \dfrac{1.32 \times 10^{18}}{3.08 \times 10^{16}}$ pc

$= 42.9$ pc

parallax angle $= \dfrac{1}{42.9}$

$= 0.023''$

MATHEMATICS – UNITS

The situation that gives rise to a change in apparent position for close stars is shown below.

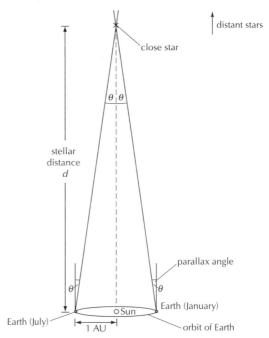

The parallax angle, θ, can be measured by observing the changes in a star's position over the period of a year. From trigonometry, if we know the distance from the Earth to the Sun, we can work out the distance from the Earth to the star, since

$$\text{Tan } \theta = \frac{\text{(distance from Earth to Sun)}}{\text{(distance from Sun to star)}}$$

Since θ is a very small angle, $\tan \theta \approx \sin \theta \approx \theta$ (in radians)

This means that $\theta \propto \dfrac{1}{\text{(distance from Earth to Star)}}$

In other words, parallax angle and distance away are inversely proportional. If we use the right units we can end up with a very simple relationship. The units are defined as follows.

The distance from the Sun to the Earth is defined to be one **astronomical unit (AU)**. It is 1.5×10^{11} m. Calculations show that a star with a parallax angle of exactly one second of arc must be 3.08×10^{16} m away (3.26 light years). This distance is defined to be one **parsec (pc)**. The name 'parsec' represents '**par**allel angle of one **sec**ond'.

If distance = 1 pc, θ = 1 second

If distance = 2 pc, θ = 0.5 second etc.

Or, distance in pc $= \dfrac{1}{\text{(parallax angle in seconds)}}$

$$d = \frac{1}{p}$$

The parallax method can be used to measure stellar distances that are less than **about 100 parsecs**. The parallax angle for stars that are at greater distances becomes too small to measure accurately. It is common, however, to continue to use the unit. The standard SI prefixes can also be used even though it is not strictly an SI unit.

1000 parsecs = 1 k pc

10^6 parsecs = 1 M pc etc.

Absolute and apparent magnitudes

THE MAGNITUDE SCALE

The everyday scale used by astronomers to compare the brightnesses of stars is the magnitude scale. The scale was introduced over 2000 years ago as a way of classifying stars. Stars were all assigned to one of six classifications according to their brightness as seen by the naked eye. Very bright stars were called magnitude 1 stars, whereas the faintest stars were called magnitude 6.

With the aid of telescopes, we can now see stars that are fainter than the magnitude 6 stars. We can also accurately measure the difference in the power being received from the stars. A magnitude 1 star is 100 times brighter than a magnitude 6 star and the scale is logarithmic. This is now used to define the magnitude scale.

The scale can seem strange at first. As the magnitude numbers get bigger and bigger, the stars are getting dimmer and dimmer. Magnitudes are negative for very bright stars.

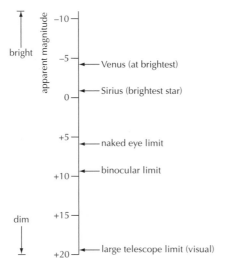

The difference between a magnitude 1 star and a magnitude 6 star is 5 'steps' on the magnitude scale and the scale is logarithmic. This means that each 'step' equates to a brightness decrease of 2.512 since

$$2.512 \times 2.512 \times 2.512 \times 2.512 \times 2.512 = 100$$

or
$$(2.512)^5 = 100$$

EXAMPLE

Use the information in the table to compare the power received from the two stars Sirius and Betelgeuse.

Difference in apparent magnitudes

$$= 0.50 - (-1.46)$$
$$= 1.96$$

$$\therefore \frac{\text{power received from Sirius}}{\text{power received from Betelgeuse}} = (2.512)^{1.96}$$
$$= 6.08$$

ABSOLUTE MAGNITUDES

As has been mentioned before, the brightness of a star depends on its luminosity **and its distance** from Earth. If two different stars have the same magnitude this does not mean they are the same size. In order to be able to compare stars in this way, the concept of **absolute magnitude** has been introduced.

The absolute magnitude of a star is the apparent magnitude that it would have IF it were observed from a distance of 10 parsecs. Since most stars are much further than 10 parsecs away, they would be brighter if observed from a distance of 10 parsecs. This means their absolute magnitudes are more negative than their apparent magnitudes.

The relationship between absolute magnitude M, apparent magnitude m and distance away d (as measured in parsecs) is given by the following formula. (This does not need to be remembered.)

$$M = m - 5 \log\left(\frac{d}{10}\right)$$

Star	Apparent magnitude m	Distance/pc	Absolute magnitude M
Sirius	−1.46	2.65	+1.4
Canopus	−0.72	70	−4.9
Alpha Centauri	−0.10	1.32	+4.3
Procyon	+0.38	3.4	+2.7
Betelgeuse	+0.50	320	−7.0

SPECTROSCOPIC PARALLAX

The term '**spectroscopic parallax**' can be confusing because it does not involve parallax at all! It is the procedure whereby the luminosity of a star can be estimated from its spectrum.

The assumption made is that the spectra from distant stars are the same as the spectra from nearby stars. If this is the case, we can use the H – R diagram for nearby stars to estimate the luminosity of the star that is further away.

Once the luminosity has been estimated, the distance to the star can be calculated from a measurement of apparent brightness. This involves using the inverse square law.

$$b = \frac{L}{4\pi d^2}$$

This procedure involves quite a lot of uncertainty. Matter between the star and the observer (for example, dust) can affect the light that is received. It would absorb some of the light and make the star's apparent brightness less than it should be. In addition, dust can scatter the different frequencies in different ways, making the identification of spectral class harder.

As the stellar distance increases, the uncertainty in the luminosity becomes greater and so the uncertainty in the distance calculation becomes greater. For this reason spectroscopic parallax is limited to measuring stellar distances up to about 10 Mpc.

Cepheid variables

PRINCIPLES

At distances greater than 10 Mpc, neither parallax nor spectroscopic parallax can be relied upon to measure the distance to a star. The essential difficulty is that when we observe the light from a star at these distances, we do not know the difference between a bright source that is far away and a dimmer source that is closer. This is the principle problem in the experimental determination of astronomical distances to other galaxies.

When we observe another galaxy, all of the stars in that galaxy are approximately the same distance away from the Earth. What we really need is a light source of known luminosity in the galaxy. If we had this then we could make comparisons with the other stars and judge their luminosities. In other words we need a 'standard candle' – that is a star of known luminosity. Cepheid variable stars provide such a 'standard candle'.

A Cepheid variable star is quite a rare type of star. Its outer layers undergo a periodic compression and contraction and this produces a periodic variation in its luminosity.

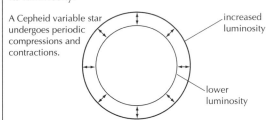

A Cepheid variable star undergoes periodic compressions and contractions.

increased luminosity

lower luminosity

These stars are useful to astronomers because the period of this variation in luminosity turns out to be related to the average absolute magnitude of the Cepheid. Thus the luminosity of a Cepheid can be calculated by observing the variations in brightness.

MATHEMATICS

The process of estimating the distance to a galaxy (in which the individual stars can be imaged) might be as follows:

- locate a Cepheid variable in the galaxy.
- measure the variation in brightness over a given period of time.
- use the luminosity-period relationship for Cepheids to estimate the average luminosity.
- use the average luminosity, the average brightness and the inverse square law to estimate the distance to the star.

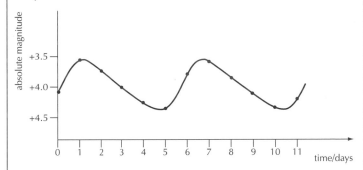

Variation of absolute magnitude for a particular Cepheid variable

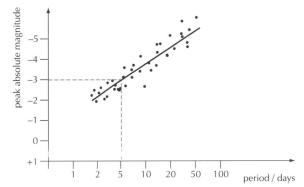

General luminosity-period graph

EXAMPLE

A cepheid variable star has a period of 5.0 days and a maximum apparent magnitude of + 8.0.

Calculate its distance away.

Using the luminosity-period graph (above)

$\Rightarrow$ absolute magnitude = – 3.0

1) Using $M = m - 5 \log \left(\dfrac{d}{10}\right)$ *or*

$$\log \frac{d}{10} = \frac{m - M}{5}$$

$$= \frac{8.0 - (-3.0)}{5}$$

$$= 2.2$$

$\therefore \dfrac{d}{10} = 10^{2.2} = 158.5$ pc

$d = 1585$ pc

2) Difference between apparent and absolute magnitude = 11.0

$\therefore \dfrac{\text{power received 10 pc away}}{\text{power received at Earth}} = (2.512)^{11}$

$$= 25130$$

$\therefore \left(\dfrac{d}{10}\right)^2 = 25130$ [inverse square law]

$$\frac{d}{10} = \sqrt{25130} = 158.5 \text{ pc}$$

$$d = 1585 \text{ pc}$$

Olbers's paradox

WHY IS IT A PARADOX?

The night sky is dark – this is an obvious experimental observation. Stars are sources of light but if we look up in a random direction we do not necessarily see a star. There seems to be nothing wrong with this but that depends on the model of the Universe that you are using.

If we look out into space, we do not see an 'edge' – it seems to carry on much the same in all directions. The constellations seem always to remain the same. From these observations Newton thought it was reasonable to assume three things about the Universe. He thought it was:
- infinite
- uniform
- static

Each of these assumptions seem reasonable enough. It is hard to imagine that the Universe is infinite, but it is worse trying to imagine that it isn't! Remember the Universe means everything – so you can't have something 'outside' the Universe.

The problem is that if the Universe really is like this then the night sky should be bright! Whatever direction you look, you should eventually come across a star. Unfortunately it is not good enough to think that the star might be so far away that it is too dim to see. If you analyze the problem in a little detail you soon find out that this explanation does not work. It really is a paradox.

MATHEMATICS

If all the stars were identical, then those at a given distance away would be the same brightness. The stars that are further away would be dimmer but **there would be more of them**.

If we start with Newton's assumptions we can imagine there to be an average number of stars in a given volume of space density ρ. Consider the stars that are at a distance between r and $r + \delta r$ – in other words the stars that are in the 'shell' of thickness δr as shown below.

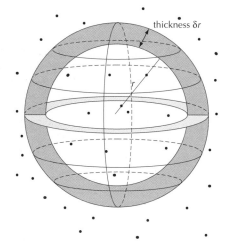

thickness δr

The volume of this shell is $4\pi r^2 \delta r$.

So the number of stars that are in this shell is $4\pi r^2 \delta r \rho$

The important thing to notice is that the number of stars in the shell $\propto r^2$.

If we move out to a shell at a greater distance, the stars in that shell will all be dimmer than the stars in the closer shell. This is due to the inverse square nature of radiation.

Brightness of one star in the shell $\propto \frac{1}{r^2}$, but the number of stars in the shell $\propto r^2$.

Overall this means that the contribution made to the overall brightness by the closer shell of stars is exactly the same as the contribution made by the more distant shell of stars.

To work out the overall brightness of the night sky, we need to add up the contributions from all the shells out from the Earth. Since the Universe is infinite the number of shells will be infinite so the night sky should be infinitely bright!

POSSIBLE SOLUTIONS TO THE PARADOX

There have been many possible attempts at explaining the paradox.

1. Perhaps the Universe is not infinite (or is non – uniform). Surprisingly a lot of people are happy to imagine that there is a 'cosmic edge' – i.e. the stars do not carry on for ever but there is a limit to the Universe. Most people who imagine this ignore the question of what is beyond the edge. This does not seem to agree with observation (essentially the Universe is the same everywhere and in all directions). The current model of the Universe is that it is infinite.

2. Perhaps the light is absorbed before it gets to us. This sounds very possible, but unfortunately if something was in the way and it absorbed the light, it would warm up. Eventually it would get hot enough to reradiate the energy.

3. Perhaps the Universe is not static (in terms of speed). The Universe is now known to be expanding – see next page. This movement means that the frequency of the light that we receive is shifted due to the Doppler effect. This shift

does not resolve the paradox, but it does mean that our model of the Universe is not correct. If it is currently expanding then what does this imply about the Universe in the past?

4. Perhaps the Universe is not static (i.e. it has changed). When we added up the contributions made by different 'shells' of stars, we assumed that the light from the far shells would add to the light from the near shells. But light has a finite velocity, so it would take time for this light to travel from the further shells. Stars are now known to have a finite lifetime – we cannot assume that they have been around forever.

On top of this, our current model of the Universe imagines that it has a limited history – it was created approximately 15 billion years ago. In this case we would only be able to see the light from stars that were less than 15 billion light years away. If we are receiving light from a finite number of stars, then the night sky will be dark.

The Big Bang model

EXPANSION OF THE UNIVERSE

If a galaxy is moving away from the Earth, the light from it would be redshifted. The surprising fact is that light from almost all galaxies shows red shifts – almost all of them are moving away from us. The Universe is expanding.

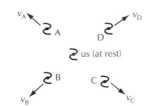

As far as we are concerned, most galaxies are moving away from us.

At first sight, this expansion seems to suggest that we are in the middle of the Universe, but this is a mistake. We only seem to be in the middle because it was we who worked out the velocities of the other galaxies. If we imagine being in a different galaxy, we would get exactly the same picture of the Universe.

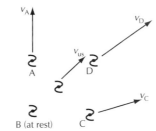

Any galaxy would see all the other galaxies moving away from it.

A good way to picture this expansion is to think of the Universe as a sheet of rubber stretching off into the distance. The galaxies are placed on this huge sheet. If the tension in the rubber is increased, everything on the sheet moves away from everything else.

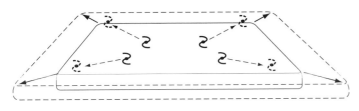

As the section of rubber sheet expands, everything moves away from everything else.

THE UNIVERSE IN THE PAST – THE BIG BANG

If the Universe is currently expanding, at some time in the past all the galaxies would have been closer together. If we examine the current expansion in detail we find that all the matter in the observable universe would have been together at the SAME point approximately 15 billion years ago.

This point, the creation of the Universe, is known as the Big Bang. It pictures all the matter in the Universe being created crushed together (very high density) and being very hot indeed. Since the Big Bang, the Universe has been expanding – which means that, on average, the temperature and density of the Universe have been decreasing. The rate of expansion has been decreasing as a result of the gravitational attraction between all the masses in the Universe.

Note that this model does not attempt to explain how the Universe was created, or by Whom. All it does is analyse what happened after this creation took place. The best way to imagine the expansion is to think of the expansion of space itself rather than the galaxies expanding into a void. The Big Bang was the creation of space and time. Einstein's theory of relativity links the measurements of space and time so properly we need to imagine the Big Bang as the creation of space **and time**. It does not make sense to ask about what happened before the Big Bang, because the notion of before and after (i.e. time itself) was created in the Big Bang.

BACKGROUND MICROWAVE RADIATION

A further piece of evidence for the Big Bang model came with the discovery of the **background microwave radiation** by Penzias and Wilson.

They discovered that microwave radiation was coming towards us from all directions in space. The strange thing was that the radiation was the same in all directions and did not seem to be linked to a source. Further analysis showed that this radiation was a very good match to theoretical black-body radiation produced by an extremely cold object – a temperature of just 3 K.

This is in perfect agreement with the predictions of Big Bang. There are two ways of understanding this.

1. All objects give out electromagnetic radiation. The frequencies can be predicted using the theoretical model of black-body radiation. The background radiation is the radiation from the Universe itself and is now cooled down to an average temperature of 3 K.

2. Some time after the Big Bang, radiation became able to travel through the Universe (see page 139 for details).

It has been travelling towards us all this time. During this time the Universe has expanded – this means that the wavelength of this radiation will have decreased (space has stretched).

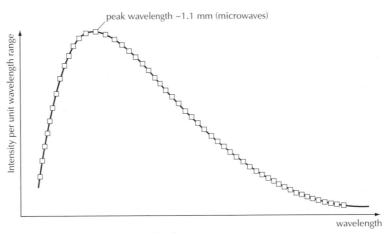

Black-body radiation for background radiation

The development of the Universe

FUTURE OF THE UNIVERSE

If the Universe is expanding at the moment, what is it going to do in the future? Remember that our current model of the Universe is that it is infinite. We can not talk about the size of the Universe, but we can talk about the size of the *observable* Universe. At the moment the furthest object that we can see is about 12 billion light years away. What is going to happen to the size of the observable Universe?

As a result of the Big Bang, other galaxies are moving away from us. If there were no forces between the galaxies, then this expansion could be thought of as being constant.

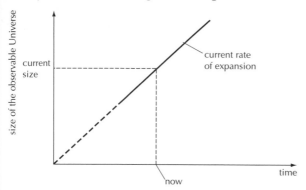

The expansion of the Universe cannot, however, have been uniform. The force of gravity acts between all masses. This means that if two masses are moving apart from one another there is a force of attraction pulling them back together. This force must have slowed the expansion down in the past. What it is going to do in the future depends on the current rate of expansion and the density of matter in the Universe.

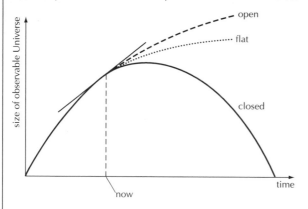

An **open Universe** is one that continues to expand forever. The force of gravity slows the rate of recession of the galaxies down a little bit but it is not strong enough to bring the expansion to a halt. This would happen if the density in the Universe were low.

A **closed Universe** is one that is brought to a stop and then collapses back on itself. The force of gravity is enough to bring the expansion to an end. This would happen if the density in the Universe were high.

A **flat Universe** is the mathematical possibility between open and closed. The force of gravity keeps on slowing the expansion down but it takes an infinite time to get to rest. This would only happen if universe were exactly the right density. One electron more, and the gravitational force would be a little bit bigger. Just enough to start the contraction and make the Universe closed.

CRITICAL DENSITY

The theoretical value of density that would create a flat Universe is called the **critical density**. Its value is not certain because the current rate of expansion is not easy to measure. It is round about 5×10^{-26} kg m^{-3} or 30 proton masses every cubic metre. If this sounds very small remember that enormous amounts of space exist that contain little or no mass at all.

The density of the Universe is not an easy quantity to measure. It is reasonably easy to estimate the mass in a galaxy by estimating the number of stars and their average mass. This calculation results in a galaxy mass which is several orders of magnitude too small. We know this because we can use the mathematics of orbital motion to work out how much mass there must be keeping the outer stars in orbit around the galactic centre.

In effect, we can see only 10% of the matter that must exist in the galaxy. This means that much of the mass of a galaxy and indeed the Universe itself must be **dark matter** – in other words we cannot observe it because it is not radiating sufficiently for us to detect it.

MACHOS, WIMPS AND OTHER THEORIES

Astrophysicists are attempting to come up with theories to explain why there is so much dark matter and what it consists of. There are a number of possible theories:

- the matter could be found in **M**assive **A**stronomical **C**ompact **H**alo **O**bjects or **MACHOs** for short. There is some evidence that lots of ordinary matter does exist in these groupings. These can be thought of as low-mass 'failed' stars or high-mass planets. They could even be black holes. These would produce little or no light.
- some fundamental particles (neutrinos) are known to exist in huge numbers. It is not known if their masses are zero or just very very small. If they turn out to be the latter then this could account for a lot of the missing mass.
- there could be new particles that we do not know about. These are the **W**eakly **I**nteracting **M**assive **P**articles. Many experimenters around the world are searching for these so-called **WIMPs**.
- perhaps our current theories of gravity are not completely correct. Some theories try to explain the missing matter as simply a failure of our current theories to take everything into account.

⒣Ⓛ Nucleosynthesis

MAIN SEQUENCE STARS

The general name for the creation of nuclei of different elements as a result of fission reactions is **nucleosynthesis**. Details of how this overall reaction takes place in the Sun do not need to be recalled.

The whole process is known as the **proton-proton cycle** or **p-p cycle**.

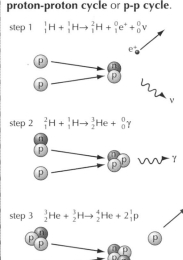

step 1 $_1^1H + _1^1H \rightarrow _1^2H + _1^0e^+ + _0^0\nu$

step 2 $_1^2H + _1^1H \rightarrow _2^3He + _0^0\gamma$

step 3 $_2^3He + _2^3H \rightarrow _2^4He + 2_1^1p$

the proton–proton cycle (p–p cycle)

In order for any of these reactions to take place, two positively charged particles (hydrogen or helium nuclei) need to come close enough for interactions to take place. Obviously they will repel one another.

This means that they must be at a high temperature.

If a large cloud of hydrogen is hot enough, then these nuclear reactions can take place spontaneously. The power radiated by the star is balanced by the power released in these reactions – the temperature is effectively constant. The star remains a stable size because the outward pressure of the radiation is balanced by the inward gravitational pull.

But how did the cloud of gas get to be at a high temperature in the first place? As the cloud comes together, the loss of gravitational potential must mean an increase in kinetic energy and hence temperature. In simple terms the gas molecules speed up as they fall in towards the centre to form a proto-star.

Once ignition has taken place, the star can remain stable for billions of years.

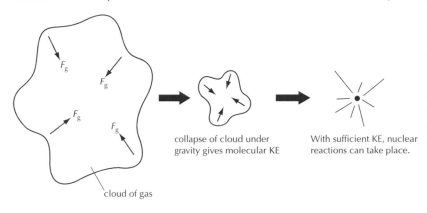

cloud of gas

collapse of cloud under gravity gives molecular KE

With sufficient KE, nuclear reactions can take place.

AFTER THE MAIN SEQUENCE

The star cannot continue in its main sequence state forever. It is fusing hydrogen into helium and at some point hydrogen in the core will become rare. The fusion reactions will happen less often. This means that the star is no longer in equilibrium and the gravitational force will, once again, cause the core to collapse.

This collapse increases the temperature of the core still further and helium fusion is now possible. The net result is for the star to increase massively in size – this expansion means that the outer layers are cooler. It becomes a red giant star.

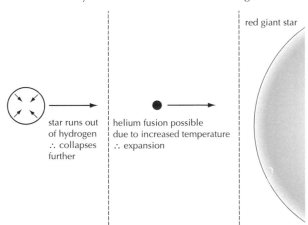

star runs out of hydrogen ∴ collapses further

helium fusion possible due to increased temperature ∴ expansion

red giant star

If it has sufficient mass, a red giant can continue to fuse higher and higher elements and the process of nucleosynthesis can continue.

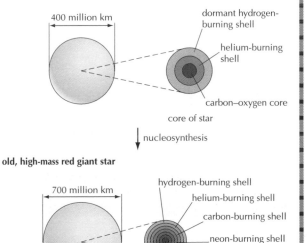

newly formed red giant star

400 million km

dormant hydrogen-burning shell

helium-burning shell

carbon–oxygen core

core of star

↓ nucleosynthesis

old, high-mass red giant star

700 million km

hydrogen-burning shell
helium-burning shell
carbon-burning shell
neon-burning shell
oxygen-burning shell
silicon-burning shell
iron core

This process of fusion as a source of energy must come to an end with the nucleosythesis of iron. The iron nucleus has the greatest binding energy per nucleon of all nuclei. In other words the fusion of iron to form a higher mass nucleus would need to take in energy rather than release energy. The star cannot continue to shine. What happens next is outlined on the following page.

POSSIBLE FATES FOR A STAR (AFTER RED GIANT PHASES)

The previous page showed that the red giant phase for a star must eventually come to an end. There are essentially two possible routes with different final states. The route that is followed depends on the initial mass of the star. An important 'critical' mass is called the **Chandrasekhar limit** and it is equal to approximately 1.4 times the mass of our Sun.

If a star has a mass less than 4 Solar masses, its remnant will be less than 1.4 Solar masses and so it is below the Chandrasekhar limit. It this case the red giant forms a **planetary nebula** and becomes a **white dwarf** which ultimately becomes invisible. The name 'planetary nebula' is another term that could cause confusion. The ejected material would not be planets in the same sense as the planets in our Solar System.

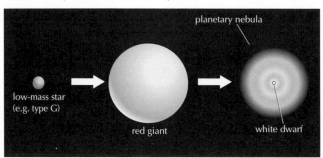

If a star is greater than 4 Solar masses, its remnant will have a mass greater than 1.4 Solar masses. It is above the Chandrasekhar limit. It this case the red giant experiences a **supernova**. It then becomes a **neutron star** or collapses to a **black hole**. See page 154. The final state again depends on mass.

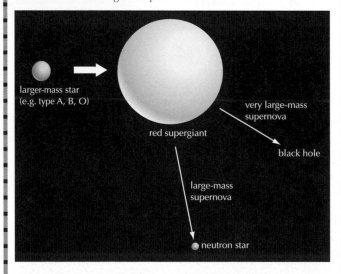

It should be emphasised that white dwarfs and neutron stars do not have a source of energy to fuel their radiation. They must be losing temperature all the time. The fact that these stars can still exist for many millions of years shows that the temperatures and masses involved are enormous.

H – R DIAGRAM INTERPRETATION

All of the possible evolutionary paths for stars that have been described here can be represented on a H – R diagram. A common mistake in examinations is for candidates to imply that a star somehow moves along the line that represents the main sequence. It does not. Once formed it stays at a stable luminosity and spectral class – i.e. it is represented by one fixed point in the H – R diagram.

evolution of a low-mass star

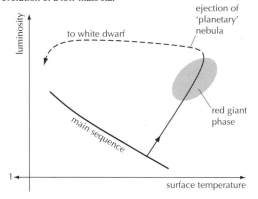

evolution of a high-mass star

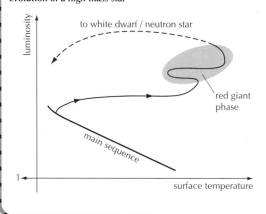

PULSARS AND QUASARS

Pulsars are cosmic sources of very weak radio wave energy that pulsate at a very rapid and precise frequency. These have now been theoretically linked to rotating neutron stars. A rotating neutron star would be expected to emit an intense beam of radio waves in a specific direction. As a result of the star's rotation, this beam moves around and causes the pulsation that we receive on Earth.

Quasi-stellar objects or quasars appear to be point-like sources of light and radio waves that are very far away. Their redshifts are very large indeed, which places them at the limits of our observations of the Universe. If they are indeed at this distance they must be emitting a great deal of power for their size (approximately 10^{40} W!). The process by which this energy is released is not well understood, but some theoretical models have been developed that rely on the existence of super-massive black holes. The energy radiated is as a result of whole stars 'falling' into the black hole.

Types of galaxy

THE MILKY WAY GALAXY

When observing the night sky a faint band of light can be seen crossing the constellations. This 'path' (or 'way') across the night sky became known as the Milky Way. What you are actually seeing is some of the millions of stars that make up our own galaxy but they are too far away to be seen as individual stars. The reason that they appear to be in a band is that our galaxy has a spiral shape.

The centre of our galaxy lies in the direction of the constellation Sagittarius. The galaxy is rotating – all the stars are orbiting each other as a result of their mutual gravitational attraction. The period of orbit is about 250 million years.

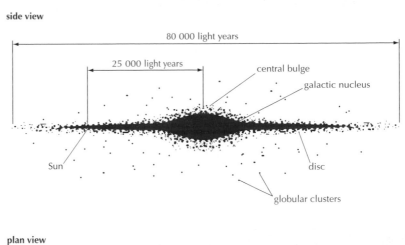

side view

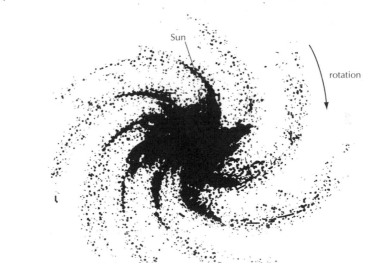

plan view

The Milky Way galaxy

OTHER GALAXIES

Other galaxies can be seen in the night sky. To the naked eye, they would appear to be 'stars' but with the aid of a telescope the structure of many can be resolved. The three main types that are observed are **spiral**, **elliptical** and **irregular**. They can be further classified into different subclasses.

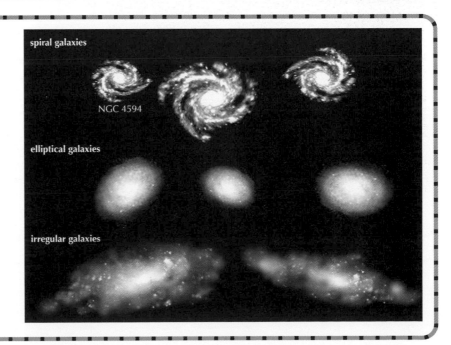

Galactic motion

DISTRIBUTIONS OF GALAXIES

Galaxies are not distributed randomly throughout space. They tend to be found clustered together. For example, in the region of the Milky Way there are twenty or so galaxies in less than 2½ million light-years.

The Virgo galactic cluster (50 million light years away from us) has over 1000 galaxies in a region 7 million light years across. On a even larger scale, the galactic clusters are grouped into huge **superclusters** of galaxies. In general, these superclusters often involve galaxies arranged together in joined 'filaments' (or bands) that are arranged as though randomly throughout empty space.

MOTION OF GALAXIES

As has been seen on page 133, it is a surprising observational fact that the vast majority of galaxies are moving away from us. The general trend is that the more distant galaxies are moving away at a greater speed as the Universe expands. This does not, however, mean that we are at the centre of the Universe – this would be observed where ever we are located in the Universe.

As explained on page 133, a good way to imagine this expansion is to think of space itself expanding. It is the expansion of space (as opposed to the motion of the galaxies through space) that results in the galaxies' relative velocities. In this model, the redshift of light can be thought of as the expansion of the wavelength due to the 'stretching' of space.

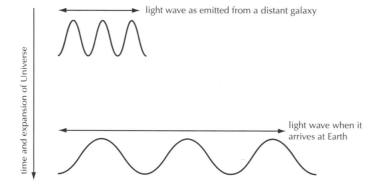

light wave as emitted from a distant galaxy

light wave when it arrives at Earth

time and expansion of Universe

MATHEMATICS

If a star or a galaxy moves away from us, then the wavelength of the light will be altered as predicted by the Doppler effect (see page 72). If a galaxy is going away from the Earth, the speed of the galaxy with respect to an observer on the Earth can be calculated from the redshift of the light from the galaxy. As long as the velocity is small when compared to the velocity of light, a simplified redshift equation can be used.

$$\frac{\Delta\lambda}{\lambda} = \frac{v}{c}$$

Where

$\Delta\lambda$ = change in wavelength of observed light (positive if wavelength is increased)

λ = wavelength of light emitted

v = relative velocity of source of light

c = speed of light

Example

A characteristic absorption line often seen in stars is due to ionized helium. It occurs at 468.6 nm. If the spectrum of a star has this line at a measured wavelength of 499.3 nm, what is the recession speed of the star?

$$\frac{\Delta\lambda}{\lambda} = \frac{(499.3 - 468.6)}{468.6}$$

$$= 6.55 \times 10^{-2}$$

$$\therefore\ v = 6.55 \times 10^{-2} \times 3 \times 10^{-8}\ \text{m s}^{-1}$$

$$= 1.97 \times 10^{7}\ \text{m s}^{-1}$$

HL Hubble's law

EXPERIMENTAL OBSERVATIONS

Although the uncertainties are large, the general trend for galaxies is that the recessional velocity is proportional to the distance away from Earth. This is Hubble's law.

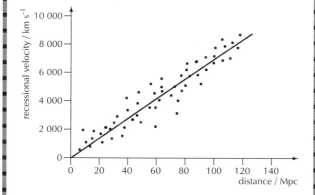

Mathematically this is expressed as

$$v \propto d$$

or

$$v = H_0 \, d$$

where H_0 is a constant known as the **Hubble constant**. The uncertainties in the data mean that the value of H_0 is not known to any degree of precision. The SI units of the Hubble constant are s^{-1}, but the unit of $km \, s^{-1} \, Mpc^{-1}$ is often used.

HISTORY OF THE UNIVERSE

If a galaxy is at a distance x, then Hubble's law predicts its velocity to be $H_0 x$. If it has been travelling at this constant speed since the beginning of the Universe, then the time that has elapsed can be calculated from

$$\text{Time} = \frac{\text{distance}}{\text{speed}}$$

$$= \frac{x}{H_0 x}$$

$$= \frac{1}{H_0}$$

This is an upper limit for the age of the Universe. The gravitational attraction between galaxies means that the speed of recession decreases all the time.

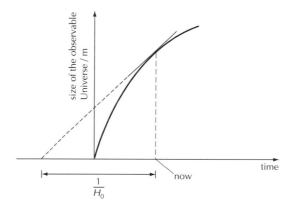

We can 'work backwards' and imagine the process that took place soon after the Big Bang.

- Very soon after the Big Bang, the Universe must have been very hot.
- As the Universe expanded it cooled. It had to cool to a certain temperature before atoms and molecules could be formed.
- The Universe underwent a short period of huge expansion that would have taken place from about 10^{-35} s after the Big Bang to 10^{-32} s.

Time	What is happening	Comments
10^{-45} s $\rightarrow$ 10^{-35} s	Unification of forces	This is the starting point
10^{-35} s $\rightarrow$ 10^{-32} s	Inflation	A rapid period of expansion – the so-called **inflationary epoch**. The reasons for this rapid expansion are not fully understood.
10^{-32} s $\rightarrow$ 10^{-5} s	Quark-lepton era	Matter and antimatter (quarks and leptons) are interacting all the time. There is slightly more matter than antimatter.
10^{-5} s $\rightarrow$ 10^{-2} s	Hadron era	At the beginning of this short period it is just cool enough for hadrons (e.g. protons and neutrons) to be stable.
10^{-2} s $\rightarrow$ 10^{2} s	Helium synthesis	During this period some of the protons and neutrons have combined to form helium nuclei. The matter that now exists is the 'small amount' that is left over when matter and antimatter have interacted.
10^{2} s $\rightarrow$ 3×10^{5} years	Plasma era (radiation era)	The formation of light nuclei has now finished and the Universe is in the form of a plasma with electrons, protons, neutrons, helium nuclei and photons all interacting.
3×10^{5} years $\rightarrow$ 10^{9} years	Formation of atoms	At the beginning of this period, the Universe has become cool enough for the first atoms to exist. Under these conditions, the photons that exist stop having to interact with the matter. It is these photons that are now being received as part of the background microwave radiation. The Universe is essentially 75% hydrogen and 25% helium.
10^{9} years $\rightarrow$ now	Formation of stars, galaxies and galactic clusters	Some of the matter can be brought together by gravitational interactions. If this matter is dense enough and hot enough, nuclear reactions can take place and stars are formed.

IB QUESTIONS – OPTION F – ASTROPHYSICS

1 The table below gives information about two nearby stars.

Star	Apparent magnitude	distance away / ly
Fomalhaut (α-Piscis Austrini)	1.2	22
Aldebaran (α-Tauri)	0.9	68

(a) To an observer on Earth which star would appear brighter? Justify your answer. [2]

(b) Explain the difference between **apparent** and **absolute** magnitudes. [2]

(c) Which star would have the *lowest numerical value* for **absolute** magnitude? Explain your answer. [2]

(d) The parallax angle for Fomalhaut is 0.148 arcseconds. Confirm that its distance away is 22 ly. [2]

(e) Would you expect Aldebaran to have a greater or smaller parallax angle than Fomalhaut? Explain your answer. [2]

Another method of determining stellar distances involves a class of variable stars called *Cepheid variables*. One of the first Cepheid variable stars to be studied is δ-Cephei. Its apparent magnitude varies over time as shown below:

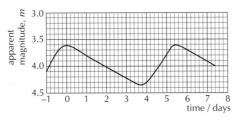

There is a relationship between peak absolute magnitude, *M*, and the time period of the variation in magnitude as shown below:

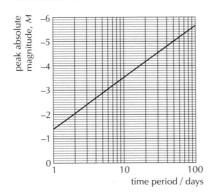

(f) Use the data above to estimate the peak **absolute** magnitude *M* for δ-Cephei. [2]

(g) The relationship between peak absolute magnitude *M*, apparent magnitude *m* and distance *d* is given by the following equation:

$$m - M = 5\log_{10}\left(\frac{d}{10pc}\right)$$

Use this equation to calculate the distance to δ-Cephei. [3]

2 (a) The spectrum of light from the Sun is shown below.

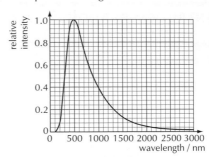

Use this spectrum to estimate the surface temperature of the Sun. [2]

(b) Outline how the following quantities can, in principle, be determined from the spectrum of a star.
  (i) The elements present in its outer layers. [2]
  (ii) Its speed relative to the Earth. [2]

3 (a) Explain how Hubble's law supports the Big Bang model of the Universe. [2]

(b) Outline **one** other piece of evidence for the model, saying how it supports the Big Bang. [3]

(c) The Andromeda galaxy is a relatively close galaxy, about 700 kps from the Milky Way, whereas the Virgo nebula is 2.3 Mps away. If Virgo is moving away at 1200 km s^{-1}, show that Hubble's law predicts that Andromeda should be moving away at roughly 400 km s^{-1}. [1]

(d) Andromeda is in fact moving **towards** the Milky Way, with a speed of about 100 km s^{-1}. How can this discrepancy from the prediction, in both magnitude and direction, be explained? [3]

(e) If light of wavelength 500 nm is emitted from Andromeda, what would be the wavelength observed from Earth? [3]

4 A star viewed from the Earth is not always a single, constant object. Many stars in the *main sequence* are, in fact, *binary stars*. For example, β-Persei is an *eclipsing binary*. Over time, stars are known to change. Some will end up as *neutron stars* or even *black holes*.

(a) What is meant by:
  (i) main sequence; [3]   **(iv)** neutron star; [1]
  (ii) binary stars; [1]   **(v)** black hole. [1]
  (iii) eclipsing binary; [1]

(b) Identify the physical processes by which a main sequence star develops into a neutron star. [4]

(c) What evidence is there for the existence of the neutron stars? [2]

(d) What property determines whether a star might develop into a neutron star or a black hole? Outline how this property can be used to predict the outcome. [2]

Frames of reference

OBSERVERS AND FRAMES OF REFERENCE

The proper treatment of large velocities involves an understanding of Einstein's theory of relativity and this means thinking about space and time in a completely different way. The reasons for this change are developed in the following pages, but they are surprisingly simple. They logically follow from two straightforward assumptions. In order to see why this is the case we need to consider what we mean by an object in motion in the first place.

A person sitting in a chair will probably think that they are at rest. Indeed from their point of view this must be true, but this is not the only way of viewing the situation. The Earth is in orbit around the Sun, so from the Sun's point of view the person sitting in the chair must be in motion. This example shows that an object's motion (or lack of it) depends on the observer.

The calculation of relative velocity was considered on page 9. This treatment, like all the mechanics in this book so far, assumes that the velocities are small enough to be able to apply Newton's laws to different frames of reference.

Is this person at rest…

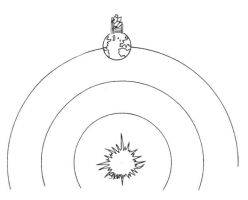

…or moving at great velocity?

GALILEAN TRANSFORMATIONS

It is possible to formalise the relationship between two different frames of reference. The idea is to use the measurement in one frame of reference to work out the measurements that would be recorded in another frame of reference. The equations that do this without taking the theory of relativity into consideration are called **Galilean transformations**.

The simplest situation to consider is two frames of reference (S and S') with one frame (S') moving past the other one (S) as shown below.

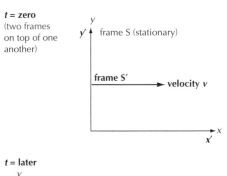

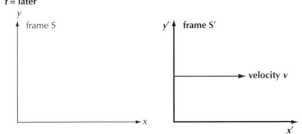

Each frame of reference can record the position and time of an event. Since the relative motion is along the x-axis, most measurements will be the same:

$$y' = y$$
$$z' = z$$
$$t' = t$$

If an event is stationary according to one frame, it will be moving according to the other frame – the frames will record different values for the x measurement. The transformation between the two is given by

$$x' = x - v\,t$$

We can use these equations to formalise the calculation of velocities. The frames will agree on any velocity measured in the y or z direction, but they will disagree on a velocity in the x-direction. Mathematically,

$$u' = u - v$$

For example, if the moving frame is going at 4 m s^{-1}, then an object moving in the same direction at a velocity of 15 m s^{-1} as recorded in the stationary frame will be measured as travelling at 11 m s^{-1} in the moving frame.

Electromagnetic theory and the speed of light

NATURE OF LIGHT

Most people know that light is an electromagnetic wave, but it is quite hard to understand what this actually means. A physical wave involves the oscillation of matter, whereas an electromagnetic wave involves the oscillation of electric and magnetic fields. The diagram below attempts to show this.

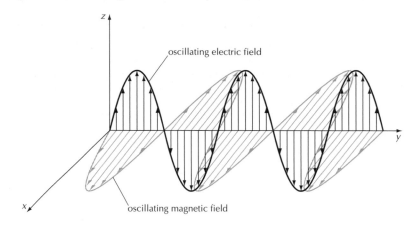

The changing electric and magnetic fields move through space – the technical way of saying this is that the fields **propagate** through space. The physics of how these fields propagate was worked out in the nineteenth century. The "rules" are summarised in four equations known as **Maxwell's equations**. The application of these equations allows the speed of all electromagnetic waves (including light) to be predicted. It turns out that this can be done in terms of the electric and magnetic constants of the medium through which they travel.

$$c = \sqrt{\frac{1}{\varepsilon_0 \mu_0}}$$

This equation does not need to be understood in detail. The only important idea is that the speed of light is **independent** of the velocity of the source of the light. In other words, a prediction from Maxwell's equations is that the speed of light in a vacuum has the same value for all observers.

FAILURE OF GALILEAN TRANSFORMATION EQUATIONS

If the speed of light has the same value for all observers then the Galilean transformation equations cannot work for light.

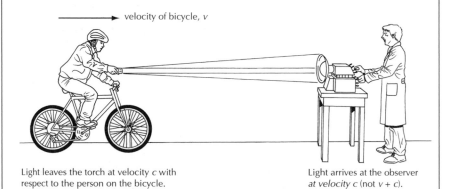

Light leaves the torch at velocity c with respect to the person on the bicycle.

Light arrives at the observer *at velocity c* (not $v + c$).

The theory of relativity attempts to work out what has gone wrong.

Concepts and postulates of special relativity

POSTULATES OF SPECIAL THEORY OF RELATIVITY

The special theory of relativity is based on two fundamental assumptions or **postulates**. If either of these postulates could be shown to be wrong, then the theory of relativity would be wrong. When discussing relativity we need to be even more than usually precise with our use of technical terms.

One important technical phrase is an **inertial frame of reference**. This means a frame of reference in which the laws of inertia (Newton's laws) apply. Newton's laws do not apply in accelerating frames of reference so an inertial frame is a frame that is either stationary or moving with constant velocity.

An important idea to grasp is that there is no fundamental difference between being stationary and moving at constant velocity. Newton's laws link forces and accelerations. If there is no resultant force on an object then its acceleration will be zero. This could mean that the object is at **rest** or it could mean that the object is **moving at constant velocity**.

The two postulates of special relativity are:
- the speed of light in a vacuum is the same constant for all inertial observers.
- the laws of physics are the same for all inertial observers.

The first postulate leads on from Maxwell's equations and can be experimentally verified. The second postulate seems completely reasonable – particularly since Newton's laws do not differentiate between being at rest and moving at constant velocity. If both are accepted as being true then we need to start thinking about space and time in a completely different way. If in doubt, we need to return to these two postulates.

SIMULTANEITY

One example of how the postulates of relativity disrupt our everyday understanding of the world around us is the concept of simultaneity. If two events happen together we say that they are **simultaneous**. We would normally expect that if two events are simultaneous to one observer, they should be simultaneous to all observers – but this is not the case! A simple way to demonstrate this is to consider an experimenter in a train.

The experimenter is positioned **exactly** in the middle of a carriage that is moving at constant velocity. She sends out two pulses of light towards the ends of the train. Mounted at the ends are mirrors that reflect the pulses back towards the observer. As far as the experimenter is concerned, the whole carriage is at rest. Since she is in the middle, the experimenter will know that:
- the pulses were sent out simultaneously.
- the pulses hit the mirrors simultaneously.
- the pulses returned simultaneously

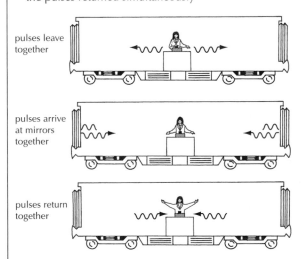

pulses leave together

pulses arrive at mirrors together

pulses return together

The situation will seem very different if watched by a stationary observer (on the platform). This observer knows that light must travel at constant speed – both beams are travelling at the same speed as far as he is concerned, so they must hit the mirrors at different times. The left-hand end of the carriage is moving towards the beam and the right hand end is moving away. This means that the reflection will happen on the left-hand end first.

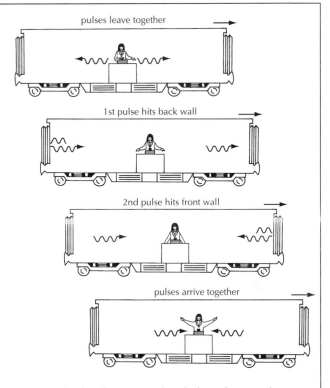

pulses leave together

1st pulse hits back wall

2nd pulse hits front wall

pulses arrive together

Interestingly, the observer on the platform does see the beams arriving back at the same time. The observer on the platform will know that:
- the pulses were sent out simultaneously.
- the left-hand pulse hit the mirror before the right hand pulse.
- the pulses returned simultaneously.

In general, simultaneous events that take place at the same point in space will be simultaneous to all observers whereas events that take place at different points in space can be simultaneous to one observer but not simultaneous to another!

Do not dismiss these ideas because the experiment seems too fanciful to be tried out. The use of a pulse of light allowed us to rely on the first postulate. This conclusion is valid whatever event is considered.

Time dilation

LIGHT CLOCK

A **light clock** is an imaginary device. A beam of light bounces between two mirrors – the time taken by the light between bounces is one 'tick' of the light clock.

As shown in the derivation the path taken by light in a light clock that is moving at constant velocity is longer. We know that the speed of light is fixed so the time between the 'ticks' on a moving clock must also be longer. This effect – that moving clocks run slow – is called **time dilation**.

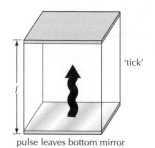

pulse leaves bottom mirror

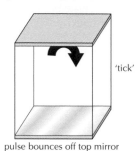

pulse bounces off top mirror

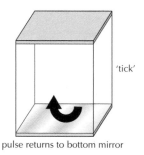

pulse returns to bottom mirror

PROPER TIME

When expressing the time taken for an event (for example the length of time that a firework is giving out light), the **proper time** is the time as measured in a frame where the event is at rest. It turns out to be the shortest possible time that any observer could correctly record for the event.

measuring how long a firework lasts

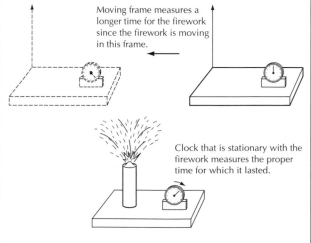

Moving frame measures a longer time for the firework since the firework is moving in this frame.

Clock that is stationary with the firework measures the proper time for which it lasted.

If A is moving past B then B will think that time is running slowly for A. From A's point of view, B is moving past A. This means that A will think that time is running slowly for B. Both views are correct!

DERIVATION OF THE EFFECT

If we imagine a stationary observer with one light clock then t is the time between 'ticks' on their stationary clock. In **this stationary frame**, a moving clock runs **slowly** and t' is the time between 'ticks' on the moving clock: t' is greater than t.

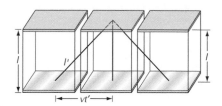

In the time t',

the clock has moved on a distance $= v\,t'$

Distance travelled by the light, $l' = \sqrt{(vt')^2 + l^2}$

$$t' = \frac{l'}{c}$$

$$= \frac{\sqrt{(vt')^2 + l^2}}{c}$$

$$\therefore \quad t'^2 = \frac{v^2 t'^2 + l^2}{c^2}$$

$$\therefore \quad t'^2\left(1 - \frac{v^2}{c^2}\right) = \frac{l^2}{c^2}$$

$$\text{but} \quad \frac{l^2}{c^2} = t^2$$

$$\therefore \quad t'^2\left(1 - \frac{v^2}{c^2}\right) = t^2$$

$$\text{or} \quad t' = \sqrt{\frac{1}{1 - \frac{v^2}{c^2}}} \times t$$

This equation is true for all measurements of time, whether they have been made using a light clock or not.

LORENTZ FACTOR

The time dilation formula expresses the time according to a stationary observer Δt in terms of the time measured on a moving clock, Δt_0

$$\Delta t = \frac{1}{\sqrt{1 - \frac{v^2}{c^2}}} \times \Delta t_0$$

We call the **Lorentz** factor, $\gamma = \dfrac{1}{\sqrt{1 - \frac{v^2}{c^2}}}$

so that $\Delta t = \gamma \Delta t_0$

At low velocities, the Lorentz factor is approximately equal to one – relativistic effects are negligible. It approaches infinity near the speed of light.

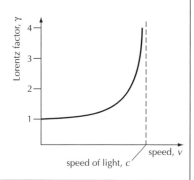

Length contraction

EFFECT

Time is not the only measurement that is affected by relative motion. There is another relativistic effect called **length contraction**. According to a (stationary) observer, the separation between two points in space contracts if there is relative motion in that direction. The contraction is in the same direction as the relative motion.

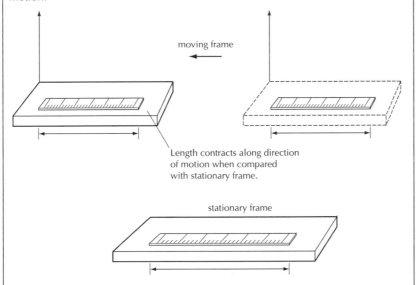

moving frame

Length contracts along direction of motion when compared with stationary frame.

stationary frame

Length contracts by the same proportion as time dilates – the Lorentz factor is once again used in the equation, but this time there is a division rather than a multiplication.

$$L = \frac{L_0}{\gamma}$$

PROPER LENGTH

As before different observers will come up with different measurements for the length of the same object depending on their relative motions. The **proper length** of an object is the length recorded in a frame where the object is at rest.

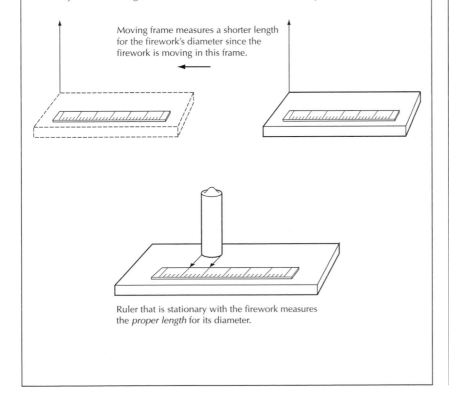

Moving frame measures a shorter length for the firework's diameter since the firework is moving in this frame.

Ruler that is stationary with the firework measures the *proper length* for its diameter.

EXAMPLE

An unstable particle has a life time of 4.0×10^{-8} s in its own rest frame. If it is moving at 98% of the speed of light calculate:

a) Its life time in the laboratory frame.

b) The length travelled in both frames.

a) $\gamma = \sqrt{\dfrac{1}{1 - (0.98)^2}}$

$= 5.025$

$\Delta t = \gamma \Delta t_0$

$= 5.025 \times 4.0 \times 10^{-18}$

$= 2.01 \times 10^{-7} \text{s}$

b) In laboratory frame, the particle moves

Length = speed × time

$= 0.98 \times 3 \times 10^8 \times 2.01 \times 10^{-7}$

$= 59.1$ m

In particle's frame, the laboratory moves

$\Delta l = \dfrac{59.1}{\gamma}$

$= 11.8$ m

(alternatively: length = speed × time

$= 0.98 \times 3 \times 10^8 \times 4.0 \times 10^{-8}$

$= 11.8$ m)

The twin paradox

As mentioned on page 144, the theory of relativity gives no preference to different inertial observers – the time dilation effect (moving clocks run slowly) is always the same. This leads to the '**twin paradox**'. In this imaginary situation, two identical twins compare their views of time. One twin remains on Earth while the other twin undergoes a very fast trip out to a distant star and back again.

As far as the twin on the Earth is concerned the other twin is a moving observer. This means that the twin that remains on the Earth will think that time has been running slowly for the other twin. When they meet up again, the returning twin should have aged less.

before   **after**

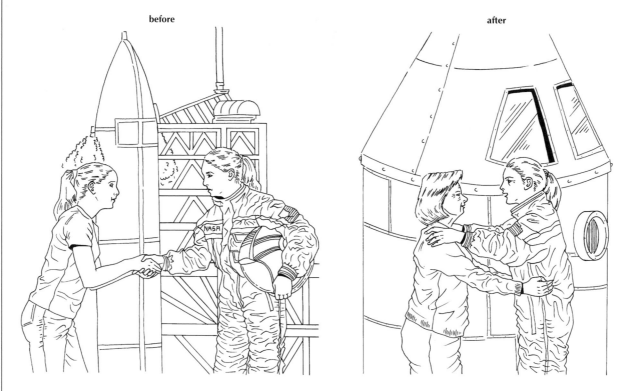

This seems a very strange prediction, but it is correct according to the time dilation formula. Remember that:
- this is a relativistic effect – time is running at different rates because of the **relative velocity** between the two twins and **not** because of the **distance** between them.
- the difference in ageing is relative. Neither twin is getting younger; as far as both of them are concerned, time has been passing at the normal rate. It's just that the moving twin thinks that she has been away for a shorter time than the time as recorded by the twin on the Earth.

The paradox is that, according to the twin who made the journey, the twin on the Earth was moving all the time and so the twin left on the Earth should have aged less. Who's version of time is correct?

The solution to the paradox comes from the realisation that the equations of special relativity are only symmetrical when the two observers are in constant relative motion. For the twins to meet back up again, one of them would have to turn around. This would involve external forces and acceleration. If this is the case then the situation is no longer symmetrical for the twins. The twin on the Earth has not accelerated so her view of the situation must be correct.

Velocity addition

MATHEMATICAL RELATIONSHIP

When two observers measure each other's velocity, they will always agree on the value. The calculation of relative velocity is not, however, normally straightforward. For example, an observer might see two objects approaching one another, as shown below.

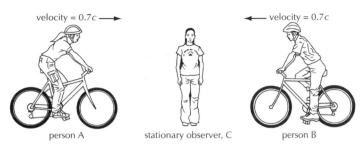

velocity = 0.7c ⟶   ⟵ velocity = 0.7c

person A   stationary observer, C   person B

If each object has a relative velocity of 0.7 c, the Galilean transformations would predict that the relative velocity between the two objects would be 1.4 c. This cannot be the case as the Lorentz factor can only be worked out for objects travelling at less than the speed of light.

The situation considered is one frame moving relative to another frame at velocity v.

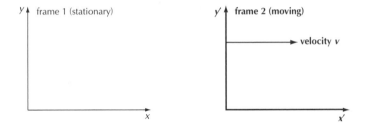

y frame 1 (stationary)

x

y' frame 2 (moving)

⟶ velocity v

x'

The equation used to move between frames is given below

$$u_x' = \frac{u_x - v}{1 - \frac{u_x v}{c^2}}$$

u_x' – the velocity in the x-direction as measured in the second frame

u_x – the velocity in the x-direction as measured in the first frame

v – the velocity of the second frame as measured in the first frame

In each of these cases, a positive velocity means motion along the positive x-direction. If something is moving in the negative x-direction then a negative velocity should be substituted into the equation.

Example

In the example above, two objects approached each other with 70% of the speed of light. So

u_x' = relative velocity of approach – to be calculated

u_x = 0.7 c

v = –0.7 c

$u_x' = \dfrac{1.4\ c}{(1 + 0.49)}$ *note the sign in the brackets*

$= \dfrac{1.4\ c}{1.49}$

$= 0.94\ c$

The top line of the relativistic addition of velocities equation can be compared with the Galilean equation for the calculation of relative velocities.

$$u_x' = u_x - v$$

At low values of v these two equations give the same value. The Galilean equation only starts to fail at high velocities.

At high velocities, the Galilean equation can give answers of greater than c, while the relativistic one always gives a relative velocity that is less than the speed of light.

Relativistic mass increase

$E = mc^2$

The most famous equation in all of physics is surely Einstein's mass-energy relationship $E = mc^2$, but where does it come from? By now it should not be a surprise that if time and length need to be viewed in a different way, then so does mass. To be absolutely precise there are two different aspects to the quantity that we call mass. In this section we shall consider what is called the **inertial mass** – (see page 17 for more details).

According to Newton's laws, a constant force produces a constant acceleration. If this was always true then any velocity at all should be achievable – even faster than light. All we have to do is apply a constant force and wait.

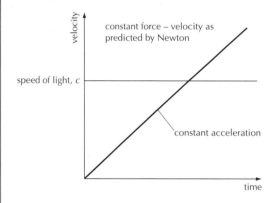

In practice, this does not happen. As soon as the speed of an object starts to approach the speed of light, the acceleration gets less and less even if the force is constant.

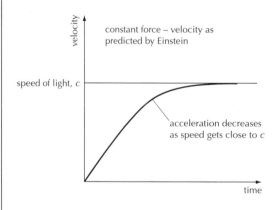

Since the force is constant and the acceleration is decreasing, the mass must be increasing. The Lorentz factor can be used to calculate the new mass. The ratio between the mass when moving with a constant velocity and the mass when the object appears at rest is equal to the Lorentz factor, γ.

$m = \gamma m_0$

Where has this extra mass come from? The force is still doing work (= force × distance). Therefore the object must be gaining in energy. The form of energy that it gains is mass-energy. Mass is just another form of energy. The equation that works out how energy and mass are related is

$$E = mc^2$$

energy in joules   mass in kilograms   speed of light in m s^{-1}

See pages 53 and 150 for more details on how to apply this equation in given situations.

REST MASS

The measurement of mass depends on relative velocity. Once again it is important to distinguish the measurement taken in the frame of the object from all other possible frames. The **rest mass** of an object is its mass as measured in a frame where the object is at rest. A frame that is moving with respect to the object would record a higher mass.

WHY NOTHING CAN GO FASTER THAN THE SPEED OF LIGHT

The relativistic mass increase formulae shows how the Lorentz factor can be used to predict the increase in mass of an object. The greater the speed of the object, the greater its mass. A greater mass means that a fixed force would produce less acceleration. If you want the acceleration to remain constant, you would need to increase the force.

As the object approaches the speed of light, its mass approaches infinite values. This means that an infinite force would be needed to accelerate it even more. Nothing can get round this absolute speed limit. The speed of light is the maximum speed that anything can ever have. Objects with mass can never actually attain this speed.

MASS-ENERGY

Mass and energy are equivalent. This means that energy can be converted into mass and vice-versa. Einstein's mass-energy equation can always be used, but one needs to be careful about how the numbers are substituted. Newtonian equations (such as $KE = \frac{1}{2} m v^2$ or momentum $= mv$) will take different forms when relativity theory is applied.

The energy needed to create a particle at rest is called the rest energy E_0 and can be calculated from the rest mass:

$E_0 = m_0 c^2$

If this particle is given a velocity, it will have a greater total energy. This means that its mass will have increased. Einstein's mass-energy equation still applies.

$E = mc^2$

but in this case the mass m is greater than the rest mass m_0 $(m = \gamma m_0)$.

EXAMPLE

An electron is accelerated through a p.d. of 1.0×10^6V. Calculate its velocity.

$$\text{Energy gained} = 1.0 \times 10^6 \times 1.6 \times 10^{-19} \text{ J}$$

$$= 1.6 \times 10^{-13} \text{ J}$$

$$E_0 = m_0 c^2 = 9.11 \times 10^{-3} \times (3 \times 10^8)$$

$$= 8.2 \times 10^{-14} \text{ J}$$

$$\therefore \text{ Total energy} = 2.42 \times 10^{-13} \text{ J}$$

$$\therefore \gamma = \frac{2.42 \times 10^{-13}}{8.2 \times 10^{-14}} = 2.95$$

$$\text{velocity} = \sqrt{1 - \frac{1}{\gamma^2}} \; c$$

$$= 0.94 \; c$$

Experiments that support special relativity

THE MUON EXPERIMENT

Muons are leptons (see page 87) – they can be thought of as a more massive version of an electron. They can be created in the laboratory but they quickly decay. Their average lifetime is 2.2×10^{-6} s as measured in the frame in which the muons are at rest.

Muons are also created high up (10 km above the surface) in the atmosphere. Cosmic rays from the Sun can cause them to be created with huge velocities – perhaps 0.99 c. As they travel towards the Earth some of them decay but there is still a detectable number of muons arriving at the surface of the Earth.

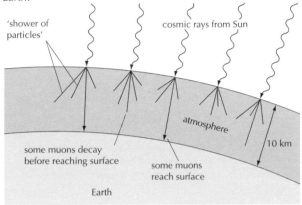

Without relativity, no muons would be expected to reach the surface at all. A particle with a lifetime of 2.2×10^{-6} s which is travelling near the speed of light (3×10^8 m s^{-1}) would be expected to travel less than a kilometre before decaying ($2.2 \times 10^{-6} \times 3 \times 10^8 = 660$ m).

The moving muons are effectively moving 'clocks'. Their high speed means that the Lorentz factor is high.

$$\gamma = \sqrt{\frac{1}{1 - 0.99^2}} = 7.1$$

Therefore an average lifetime of 2.2×10^{-6} s in the muons' frame of reference will be time dilated to a longer time as far as a stationary observer on the Earth is concerned. From this frame of reference they will last, on average, 7.1 times longer. Many muons will still decay but some will make it through to the surface – this is exactly what is observed.

In the muons' frame they exist for 2.2×10^{-6} s on average. They make it down to the surface because the atmosphere (and the Earth) is moving with respect to the muons. This means that the atmosphere will be length-contracted. The 10km distance as measured by an observer on the Earth will only be $\frac{10}{7.1} = 1.4$ km. A significant number of muons will exist long enough for the Earth to travel this distance.

THE MICHELSON-MORLEY EXPERIMENT

With hindsight, the **Michelson-Morley experiment** provides crucial support for the first postulate of special relativity – the constancy of the speed of light. The aim of the experiment was to measure the speed of the Earth through space. Using the technical language of the time, they were trying to measure the speed of the Earth through the Aether.

The experiment involved two beams of light travelling down two paths at right angles to one another. Having travelled different paths, the light was brought together where it interfered and produced fringes of constructive and destructive interference.

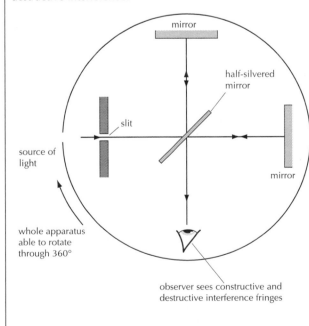

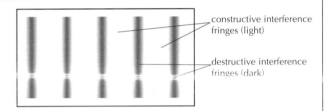

From the Galilean point of view, the motion of the Earth through space would affect the time of travel for the two beams by different amounts. Suppose the Earth was moving through the Aether in a direction that was parallel to one arm, the speed of light down that beam would be ($c + v$) and on the way back the speed would be ($c - v$). This is different to the speed of light in the arm that is at right angles to the motion. Here the light would have a speed of $\sqrt{(c^2 + v^2)}$.

If the whole apparatus were rotated around, the speed down the paths would change. This would move the interference pattern. The idea was to measure the change and thus work out the speed of the Earth through the Aether.

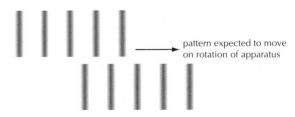

The experiment was tried but the rotation of the apparatus did not produce any observable change in the interference pattern. This null result can be easily understood from the first postulate of relativity – the constancy of the speed of light. The interference pattern does not change because the speed of light along the paths is always the same. It is unaffected by the motion of the Earth.

(HL) Relativistic momentum and energy

EQUATIONS

The connection between mass and energy was introduced on page 148:

$$E_0 = m_0 c^2$$

$$E = mc^2$$

Other equations can be derived for relativistic situations. Physical laws that apply in Newtonian situations also apply in relativistic situations. However the concepts often have to be refined to take into account the new ways of viewing space and time.

For example, in Newtonian mechanics, momentum p is defined as the product of mass and velocity.

$$p = mv$$

In relativity it has a similar form, but the Lorentz factor needs to be taken into consideration.

$$p = \gamma m_0 v$$

The momentum of an object is related to its total energy. In relativistic mechanics, the relationship can be stated as

$$E^2 = p^2 c^2 + m_0^2 c^4$$

In Newtonian mechanics, the relationship between energy and momentum is

$$E = \frac{p^2}{2m}$$

Do not be tempted to use the standard Newtonian equations – if the situation is relativistic, then you need to use the relativistic equations.

UNITS

SI units can be applied in these equations. Sometimes, however, it is useful to use other units instead.

At the atomic scale, the joule is a huge unit. Often the electronvolt (eV) is used. One electronvolt is the energy gained by one electron if it moves through a potential difference of 1 volt. Since

$$\text{Potential difference} = \frac{\text{energy difference}}{\text{change}}$$

$$1\text{ eV} = 1\text{ V} \times 1.6 \times 10^{-19}\text{ C}$$

$$= 1.6 \times 10^{-19}\text{ J}$$

In fact the electronvolt is too small a unit, so the standard SI multiples are used

$$1\text{ keV} = 1000\text{ eV}$$

$$1\text{ MeV} = 10^6\text{ eV}\quad\text{etc.}$$

Since mass and energy are equivalent, it makes sense to have comparable units for mass. The equation that links the two ($E = mc^2$) defines a new unit for mass – the MeV c^{-2}. The speed of light is included in the unit so that no change of number is needed when switching between mass and energy – If a particle of mass of 5 MeV c^{-2} is converted completely into energy, the energy released would be 5 MeV. It would also be possible to used keV c^{-2} or GeV c^{-2} as a unit for mass.

In a similar way, the easiest unit for momentum is the MeV c^{-1}. This is the best unit to use if using the equation which links relativistic energy and momentum.

EXAMPLE

The Large Electron / Positron (LEP) collider at the European Centre for Nuclear Research (CERN) accelerates electrons to total energies of about 90 GeV. These electrons then collide with *positrons* moving in the opposite direction as shown below. Positrons are identical in rest mass to electrons but carry a positive charge. The positrons have the same energy as the electrons.

Electron	Positron
• ———→	←——— •
Total energy = 90 GeV	Total energy = 90 GeV

(a) Use the equations of special relativity to calculate,

    (i) the velocity of an electron (with respect to the laboratory);

> **Total energy = 90 GeV = 90,000 MeV**

> **Rest mass = 0.5 MeVc^{-2} ∴ γ = 18000 (huge)**

> $$\therefore v \simeq c$$

    (ii) the momentum of an electron (with respect to the laboratory).

> $$p^2 c^2 = E^2 - m_0^2 c^4$$

> $$\simeq E^2$$

> $$p \simeq 90\text{ GeV}c^{-1}$$

(b) For these two particles, estimate their relative velocity of approach.

> **since γ so large**

> **relative velocity $\simeq c$**

(c) What is the total momentum of the system (the two particles) before the collision?

> **zero**

(d) The collision causes new particles to be created.

    (i) Estimate the maximum total rest mass possible for the new particles.

> **Total energy available = 180 GeV**

> **∴ max total rest mass possible = 180 GeVc^{-2}**

    (ii) Give one reason why your answer is a *maximum*.

> **Above assumes that particles were created at rest**

General relativity – The equivalence principle

GRAVITATIONAL AND INERTIAL MASS

General relativity attempts to include gravitational effects in the ideas developed by the special theory of relativity.

The mass of an object is a single quantity but it can be thought of as being two very different properties. In order to distinguish between the two they are sometimes given different names – **gravitational mass** and **inertial mass**.

The more gravitational mass that an object has, the greater the gravitational force of attraction between it and every other mass. See page 61.

Inertial mass is concerned with how the object moves if acted upon by any force – the nature of the force does not matter.

It is a surprising experimental fact that both of these different aspects are explained by the one quantity that we call mass. For all objects, gravitational mass and inertial mass are equivalent. We do not know why this should be the case.

PRINCIPLE OF EQUIVALENCE

One of Einstein's 'thought experiments' considers how an observer's view of the world would change if they were accelerating. The example below considers an observer inside a closed spaceship.

There are two possible situations to compare.
- the rocket could be far away from any planet but accelerating forwards.
- the rocket at rest on the surface of a planet.

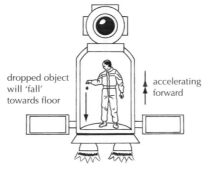

dropped object will 'fall' towards floor — accelerating forward

dropped object will fall towards floor — rocket at rest on planet

planet

astronaut feels a force when rocket is accelerating forward

astronaut feels a force when rocket is at rest on the surface of a planet

Although these situations seem completely different, the observer **inside** the rocket would interpret these situations as being identical.

This is Einstein's 'principle of equivalence' – a postulate that states that there is no difference between an accelerating frame of reference and a gravitational field.

BENDING OF LIGHT

Einstein's principle of equivalence suggests that a gravitational field should bend light rays! There is a small window high up in the rocket that allows a beam of light to enter.

In both of the cases to the right, the observer is an **inertial** observer and would see the light shining on the wall at the point that is exactly opposite the small window. If, however, the rocket was accelerating upwards (see below) then the beam of light would hit a point on the wall **below** the point that is opposite the small window.

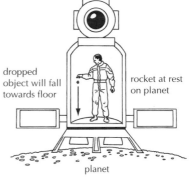

rocket at rest in space

window — light hits wall opposite window

rocket moving with constant velocity

final position of window when light hits

light hits wall opposite window

original position of window

rocket moves upwards at constant velocity

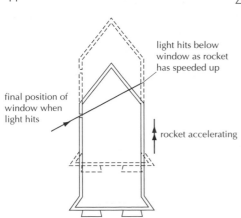

final position of window when light hits

light hits below window as rocket has speeded up

rocket accelerating

But Einstein's principle of equivalence states that there is no difference between an accelerating observer and inertial observer in a gravitational field. If this is true then light should follow a curved path in a gravitational field. This effect does happen!

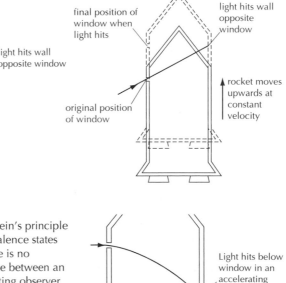

Light hits below window in an accelerating rocket *and* in a stationary rocket in a gravitational field.

view inside rocket

HL Space-time

SPACE-TIME

Relativity has shown that our Newtonian ideas of space and time are incorrect. Two inertial observers will generally disagree on their measurements of space and time but they will agree on a measurement of the speed of light. Is there anything else upon which they will agree?

In relativity, a good way of imagining what is going on is to consider everything as different 'events' in something called space-time. From one observer's point of view, three co-ordinates (x, y and z) can define a position in space. One further 'coordinate' is required to define its position in time (t). An event is a given point specified by these four co-ordinates (x, y, z, t).

As a result of length contraction and time dilation, another observer would be expected to come up with totally different numbers for all of these four measurements – (x^1, y^1, z^1, t^1). The amazing thing is that these two observers will agree on something. This is best stated mathematically:

$$x^2 + y^2 + z^2 - c^2 t^2 = (x^1)^2 + (y^1)^2 + (z^1)^2 - c^2(t^1)^2$$

On normal axes, Pythagoras's theorem shows us that the quantity $\sqrt{(x^2 + y^2 + z^2)}$ is equal to the length of the line from the origin, so $(x^2 + y^2 + z^2)$ is equal to (the length of the line) ^{2}. In other words, it is the separation in space.

$$(\text{Separation in space})^2 = (x^2 + y^2 + z^2)$$

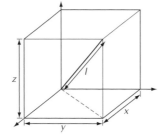

$$l^2 = x^2 + y^2 + z^2$$

The two observers agree about something very similar to this, but it includes a co-ordinate of time. This can be thought of as the separation in imaginary four-dimensional space-time.

$$(\text{Separation in space-time})^2 = (x^2 + y^2 + z^2 - c^2 t^2)$$

We cannot represent all four dimensions on the one diagram, so we usually limit the number of dimensions of space that we represent. The simplest representation has only one dimension of space and one of time as shown below.

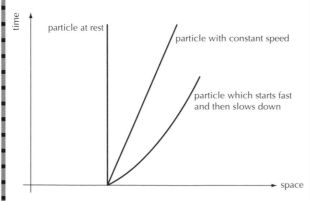

An object (moving or stationary) is always represented as a line in space-time.

EFFECT OF GRAVITY ON SPACE-TIME

The Newtonian way of describing gravity is in terms of the forces between two masses. In general relativity the way of thinking about gravity is not to think of it as a force, but as changes in the shape (warping) of space-time. The warping of space-time is caused by mass. Think about two travellers who both set off from different points on the Earth's equator and travel north.

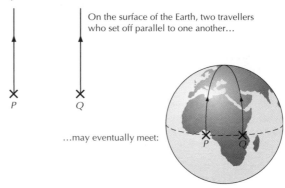

On the surface of the Earth, two travellers who set off parallel to one another…

…may eventually meet:

As they travel north they will get closer and closer together. They could explain this coming together in terms of a force of attraction between them or they could explain it as a consequence of the surface of the Earth being curved. The travellers have to move in straight lines across the surface of the Earth so their paths come together.

Einstein showed how space-time could be thought of as being curved by mass. The more matter you have, the more curved space-time becomes. Moving objects follow the curvature of space-time or in other words, they take the shortest path in space-time. As has been explained, it is very hard to imagine the four dimensions of space-time. It is easier to picture what is going on by representing space-time as a flat two-dimensional sheet.

space–time represented by flat sheet

Any mass present warps (or bends) space-time. The more mass you have the greater the warping that takes place. This warping of space can be used to describe the the orbit of the Earth around the Sun. The diagram below represents how Einstein would explain the situation. The Sun warps space-time around itself. The Earth orbits the Sun because it is travelling along the shortest possible path in space-time. This turns out to be a curved path.

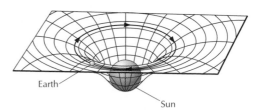

Earth

Sun

- mass 'tells' space how to curve.
- space 'tells' matter how to move.

HL Gravitational red shift

CONCEPT

The general theory of relativity makes other predictions that can be experimentally tested. One such effect is **gravitational redshift** – clocks slow down in a gravitational field. In other words a clock on the ground floor of a building will run slowly when compared with a clock in the attic – the attic is further away from the centre of the Earth.

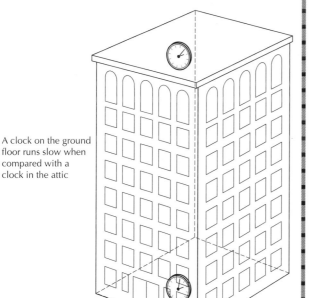

A clock on the ground floor runs slow when compared with a clock in the attic

The same effect can be imagined in a different way. We have seen that a gravitational field affects light. If light is shone away from a mass (for example the Sun), the photons of light must be increasing their gravitational potential energy as they move away. This means that they must be decreasing their total energy. Since frequency is a measure of the energy of a photon, the observed frequency away from the source must be less than the emitted frequency.

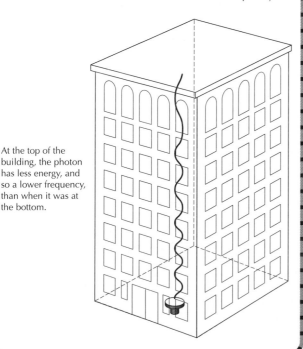

At the top of the building, the photon has less energy, and so a lower frequency, than when it was at the bottom.

MATHEMATICS

This gravitational time dilation effect can be mathematically worked out for a uniform gravitational field g. The change in frequency Δf is given by

$$\frac{\Delta f}{f} = \frac{g\Delta h}{c^2}$$

where

f – is the frequency emitted at the source

g – is the gravitational field strength (assumed to be constant)

Δh – is the height difference and

c – is the speed of light.

EXAMPLE

A UFO travels at such a speed to remain above one point on the Earth at a height of 200 km above the Earth's surface. A radio signal of frequency of 1.10 MHz is sent to the UFO.

(i) What is the frequency received by the UFO?

(ii) If the signal was reflected back to Earth, what would be the observer frequency of the return signal? Explain your answer.

(i) $\quad f = 1.1 \times 10^8$ Hz

$\quad g = 10$ m s^{-2}

$\quad \Delta h = 2.0 \times 10^5$ m

$\therefore \ \Delta f = \dfrac{10 \times 2.0 \times 10^5}{(3 \times 10^8)} \times 1.1 \times 10^8$ Hz

$\quad = 2.4 \times 10^{-3}$ Hz

$\therefore \ f$ received $= 1.1 \times 10^8 - 2.4 \times 10^{-8}$

$\quad\quad\quad = 109999999.998$ Hz

$\quad\quad\quad = 1.1 \times 10^8$ Hz

(ii) The return signal will be gravitationally blueshifted. Therefore it will arrive back at **exactly** the same frequency as emitted.

HL Black holes

DESCRIPTION

When a star has used up all of its nuclear fuel, the force of gravity makes it collapse down on itself (see the astrophysics section for more details). The more it contracts the greater the density of matter and thus the greater the gravitational field near the collapsing star. In terms of general relativity, this would be described in terms of the space near a collapsing star becoming more and more curved. The curvature of space becomes more and more severe depending on the mass of the collapsing star.

If the collapsing star is less than about 1.4 times the mass of the Sun, then the electrons play an important part in eventually stopping this contraction. The star that is left is called a white dwarf. If the collapsing star is greater than this, the electrons cannot halt the contraction. A contracting mass of up to three times the mass of the Sun can also be stopped – this time the neutrons play an important role and the star that is left is called a neutron star. The curvature of space-time near a neutron star is more extreme than the curvature near a white dwarf.

At masses greater than this we do not know of any process that can stop the contraction. Space-time around the mass becomes more and more warped until eventually it becomes so great that it folds in over itself. What is left is called a black hole. All the mass is concentrated into a point – the **singularity**.

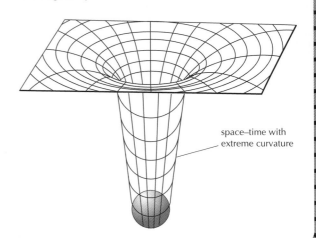

space–time with extreme curvature

SCHWARZCHILD RADIUS

The curvature of space-time near a black hole is so extreme that nothing, not even light, can escape. Matter can be attracted into the hole, but nothing can get out since nothing can travel faster than light. The gravitational forces are so extreme that light would be severely deflected near a black hole.

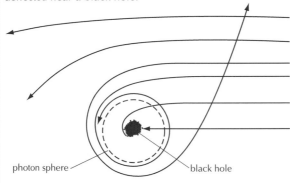

photon sphere   black hole

If you were to approach a black hole, the gravitational forces on you would increase. The first thing of interest would be the **photon sphere**. This consists of a very thin shell of light photons captured in orbit around the black hole. As we fall further in the gravitational forces increase and so the escape velocity at that distance also increases.

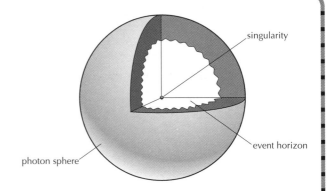

singularity

photon sphere

event horizon

At a particular distance from the centre, called the **Schwarzchild radius**, we get to a point where the escape velocity is equal to the speed of light. Newtonian mechanics predicts that the escape velocity v from a mass M of radius r is given by the formula

$$v = \sqrt{\frac{2GM}{r}}$$

If the escape velocity is the speed of light, c, then the Schwarzchild radius would be given by

$$R_{Sch} = \frac{2GM}{c^2}$$

It turns out that this equation is also correct if we use the proper equations of general relativity. If we cross the Schwarzchild radius and get closer to the singularity, we would no longer be able to communicate with the Universe outside. For this reason crossing the Schwarzchild radius is sometimes called crossing the **event horizon**.

EXAMPLE

Calculate the size of a black hole that has the same mass as our sun (1.99×10^{30} kg).

$$R_{Sch} = \frac{2 \times 6.67 \times 10^{-11} \times 1.99 \, 10^{30}}{(3 \times 10^8)^2}$$

$$= 2949.6 \text{ m}$$

$$= 2.9 \text{ km}$$

EVIDENCE TO SUPPORT GENERAL RELATIVITY

Bending of star light

The predictions of general relativity, just like those of special relativity, seem so strange that we need strong experimental evidence. One main prediction was the bending of light by a gravitational field. One of the first experiments to check this effect was done by a physicist called Arthur Eddington in 1919.

The idea behind the experiment was to measure the deflection of light (from a star) as a result of the Sun's mass. During the day, the stars are not visible because the Sun is so bright. During a solar eclipse, however, stars are visible during the few minutes when the Moon blocks all of the light from the Sun. If the positions of the stars during the total eclipse were compared with the positions of the same stars recorded at different time, the stars that appeared near the edge of the Sun would have moved.

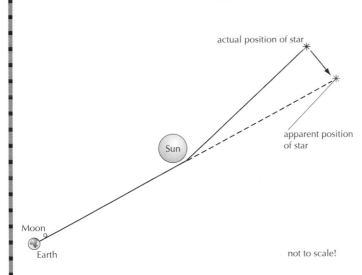

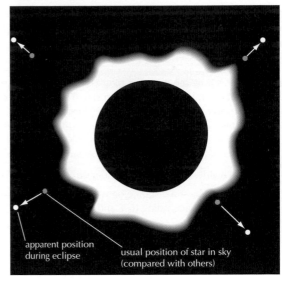

The angle of the shift of these stars turned out to be exactly the angle as predicted by Einstein's general theory of relativity.

Gravitational lensing

The bending of the path of light or the warping of space-time (depending on which description you prefer) can also produce some very extreme effects. Massive galaxies can deflect the light from quasars (or other very distance sources of light) so that the rays bend around the galaxy as shown below.

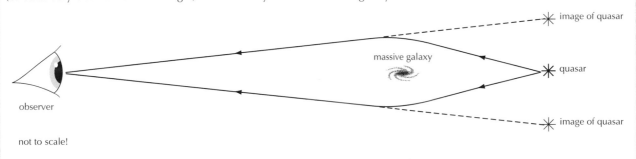

In this strange situation, the galaxy is acting like a lens and we can observe multiple images of the distant quasar.

EVIDENCE TO SUPPORT GRAVITATIONAL REDSHIFT

Pound-Rebka

The decrease in the frequency of a photon as it climbs out of a gravitational field can be measured in the laboratory. The measurements need to be very sensitive, but they have been successfully achieved on many occasions. One of the experiments to do this was done in 1960 and is called the **Pound-Rebka** experiment. The frequencies of gamma ray photons were measured after they ascended or descended Jefferson Physical Laboratory Tower at Harvard University.

Atomic clock frequency shift

Because they are so sensitive, comparing the difference in time recorded by two identical atomic clocks can provide a direct measurement of gravitational redshift. One of the clocks is taken to high altitude by a rocket, whereas a second one remains on the ground. The clock that is at the higher altitude will run faster.

IB QUESTIONS – OPTION G – RELATIVITY

1 *Relativity and simultaneity*

(a) State two postulates of the special theory of relativity. [2]

Einstein proposed a 'thought experiment' along the following lines. Imagine a train of proper length 100 m passing through a station at half the speed of light. There are two lightning strikes, one at the front and one at the rear of the train, leaving scorch marks on both the train and the station platform. Observer S is standing on the station platform midway between the two strikes, while observer T is sitting in the middle of the train. Light from each strike travels to both observers.

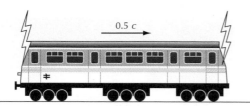

(b) If observer S on the station concludes from his observations that the two lightning strikes occurred simultaneously, explain why observer T on the train will conclude that they did **not** occur simultaneously. [4]

(c) Which strike will T conclude occurred first? [1]

(d) What will be the distance between the scorch marks on the *train*, according to T and according to S? [3]

(e) What will be the distance between the scorch marks on the *platform*, according to T and according to S? [2]

2 An electron is travelling at a constant speed in a vacuum. A laboratory observer measures its speed as 95% of the speed of light and the length of its journey to be 100 m.

(a) Show that for these electrons, $\gamma = 3.2$.

(b) What is the length of the journey in the **electron's** frame of reference? [1]

(c) What is the time taken for this journey in the **electron's** frame of reference? [2]

(d) What is the mass of the electron according to the **laboratory** observer? [2]

(e) Use the axes below to show how the observed mass of the electron will change with velocity as measured by the laboratory observer. There is no need to do any further calculations. [3]

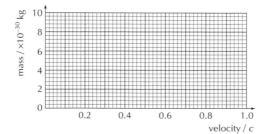

3 In a laboratory experiment two identical particles (P and Q), each of rest mass m_0, collide. In the **laboratory frame of reference**, they are both moving at a velocity of 2/3 c. The situation before the collision is shown in the diagram below.

Before:

(a) In the laboratory frame of reference,
  (i) what is the **total momentum** of P and Q? [1]
  (ii) what is the **total energy** of P and Q? [3]

The same collision can be viewed according to **P's frame of reference** as shown in the diagrams below.

(b) In P's frame of reference,
  (i) what is Q's velocity, *v*? [3]
  (ii) what is the **total momentum** of P and Q? [3]
  (iii) what is the **total energy** of P and Q? [3]

(c) As a result of the collision, many particles and photons are formed, but the total energy of the particles depends on the frame of reference. Do the observers in each frame of reference agree or disagree on the number of particles and photons formed in the collision? Explain your answer. [2]

4 Two space travellers Lee and Anna are put into a state of hibernation in a ventilated capsule in a spaceship, for a long trip to find another habitable planet. They eventually awake, but do not know whether the ship is still travelling or whether they have landed. They feel attracted toward the floor of the capsule, an experience rather like weak gravity. Lee says the spaceship must have landed on a planet and they are experiencing its gravitational attraction. Anna says the spaceship must be accelerating and the capsule floor is pushing on them.

(a) Hoping to decide which of them is right, they try an experiment. They release a hammer in mid air, and it accelerates straight to the floor. Does this observation help them decide? How would *each* of them explain the motion of the hammer? [4]

(b) They propose another experiment, namely to shine monochromatic light from the floor to the ceiling of the capsule and use sensitive apparatus to detect any change in frequency.
  (i) How would Lee explain how a redshift arises, viewing the radiation as photons moving in a gravitational field? [2]
  (ii) How would Anna explain how a redshift arises, viewing the radiation in terms of wavefronts arriving at a detector whose speed is increasing? [2]

(c) Can Lee and Anna perform *any* experiment in the capsule which could distinguish whether they are on the surface of a planet or accelerating in space? State why. [1]

(d) Later they notice that the gravitational-like sensation starts diminishing gradually, until they eventually 'float weightless' in the capsule. Lee suggests that they must have taken off from the planet, and as they got further away its gravitational attraction diminished until it was negligible. Anna suggests that the spaceship must have gradually reduced its thrust and acceleration to zero. Which explanation is feasible, or is there no way to tell who is right? Explain. [3]

Speed of light

ELECTROMAGNETIC WAVES

Visible light is one part of a much larger spectrum of similar waves that are all electromagnetic.

Charges that are accelerating generate electromagnetic fields. If an electric charge oscillates, it will produce a varying electric and magnetic field at right angles to one another.

These oscillating fields propagate (move) as a transverse wave through space. Since no physical matter is involved in this propagation, they can travel through a vacuum. The speed of this wave can be calculated from basic electric and magnetic constants and it is the same for all electromagnetic waves, 3.0×10^8 m s^{-1}. See page 142.

Although all electromagnetic waves are identical in their nature, they have very different properties. This is because of the huge range of frequencies (and thus energies) involved in the electromagnetic spectrum.

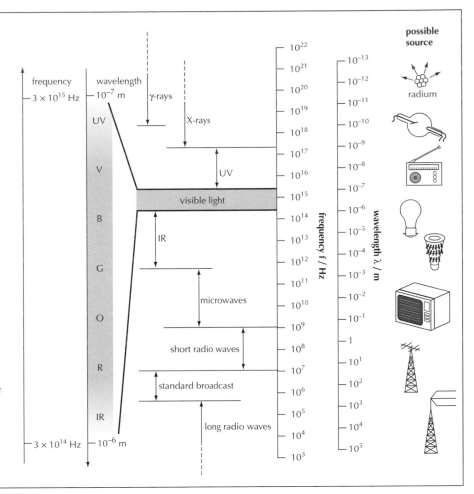

EXPERIMENTAL MEASUREMENT

The speed of light is so high that special techniques are needed to measure it. Michelson's method involved bouncing light between two mountains that were 34 km apart. The set-up is shown here.

- during the experiment, the octagonal mirror is made to rotate.
- at most frequencies of rotation, no image will be seen in the eyepiece.
- if the rotational frequency is a certain value, the mirror will make exactly an eighth of a revolution in the time it takes for the light to go between the mountain tops. This means that an image will be seen.

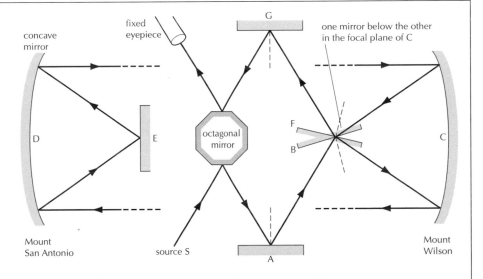

- the speed of the light can then be calculated using

$$\text{speed} = \frac{\text{distance travelled}}{\text{time taken}} = \frac{\text{twice the distance between the mountains}}{\text{time for mirror to make } \frac{1}{8} \text{ revolution}}$$

At the time of this experiment (the 1920s), the speed of light was a **measured quantity**. The experimental value was quoted in terms of two **defined units** (the metre and the second). Nowadays the speed of light in a vacuum is defined to be a certain value and the metre is defined as the distance travelled by light in $\frac{1}{299792458}$ seconds.

Dispersion

DISPERSION

White light is, in fact, a combination of all the visible frequencies. These different frequencies of light are perceived as different colours of the rainbow. The splitting up of white light into its component colours is called **dispersion**. A prism causes the dispersion of light because the refractive indexes are slightly different for each of the different colours.

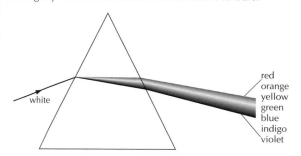

Note that
- the refraction takes place at both surfaces.
- red light is bent the least and blue light is bent the most. In other words the refractive index for red light is smaller than for blue.

It is also possible to do the same process in reverse. An addition of all the different frequencies of light would result in white light. The human eye and brain system also perceive different combinations of light as a single colour.

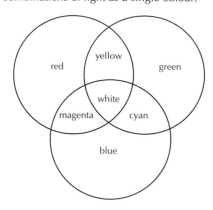

Note that
- the addition of three colours (red, green and blue) is used in television sets to create all the colours in a picture.
- the above diagram represent overlapping circles **of light**. This is not the same as using overlapping filters or mixing paints.

Lasers

MONOCHROMATIC

Laser light is **monochromatic**. It contains only a very very narrow band of frequencies. This is very different to the light given out by a light bulb, which contains a mixture of many different frequencies.

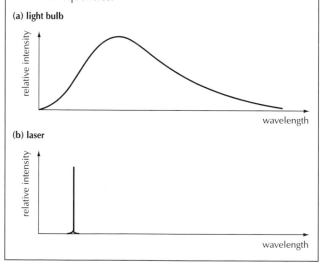

COHERENT

Laser light is **coherent**. The oscillations that make up the waves are linked together. Light is always emitted in 'packets' of energy called photons. Each photon in laser light is in phase with all the other photons that are emitted. This is very different to the light given out by a light bulb. Each atom that emits light acts independently from the other atoms. This means that every photon has an independent phase.

(a) light bulb   **(b) laser**

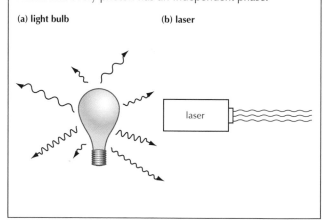

APPLICATIONS

The are many uses of laser in technology, industry and medicine. Some of these rely on the fact that laser light does not tend to diverge. This means that lasers can provide energy that is concentrated in a small region. When used medically, the energy supplied by the laser can be enough to heat and destroy a small area of tissue while leaving neighbouring areas virtually unaffected. A laser used in this way to cut through tissue is called laser surgery. It has the additional advantage that the blood vessels that are cut by the process also tend to be sealed at the same time. This means that there is less bleeding than when using a knife.

Other possible applications include:
- technology (bar-code scanners, laser discs)
- industry (surveying, welding and machining metals)
- medicine (destroying tissue in small areas, attaching the retina, corneal correction)

Nature of reflection

TYPES OF REFLECTION

When a single ray of light strikes a smooth mirror it produces a single reflected ray. This type of 'perfect' reflection is very different to the reflection that takes place from an uneven surface such as the walls of a room. In this situation, a single incident ray is generally scattered in all directions. This is an example of a **diffuse** reflection.

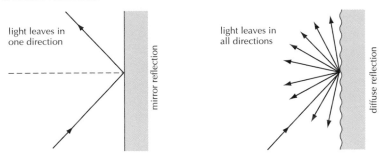

We see objects by receiving light that has come from them. Most objects do not give out light by themselves so we cannot see them in the dark. Objects become visible with a source of light (e.g. the Sun or a light bulb) because diffuse reflections have taken place that scatter light from the source towards our eyes.

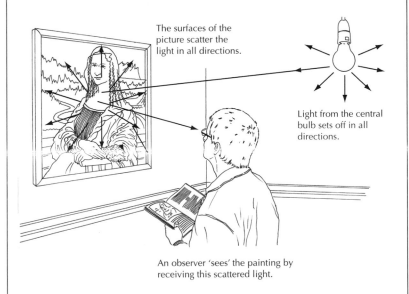

The surfaces of the picture scatter the light in all directions.

Light from the central bulb sets off in all directions.

An observer 'sees' the painting by receiving this scattered light.

Our brains are able to work out the location of the object by assuming that rays travel in straight lines.

LAW OF REFLECTION

The location and nature of optical images can be worked out using **ray diagrams** and the principles of **geometric optics**. A ray is a line showing the direction in which light energy is propagated. The ray must always be at right angles to the wave front. The study of geometric optics ignores the wave and particle nature of light.

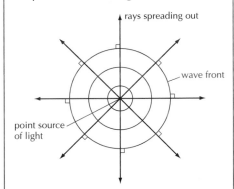

When a mirror reflection takes place, the direction of the reflected ray can be predicted using the laws of reflection. In order to specify the ray directions involved, it is usual to measure all angles with respect to an imaginary construction line called the **normal**. For example, the incident angle is always taken as the angle between the incident ray and the normal. The normal to a surface is the line at right angles to the surface as shown below.

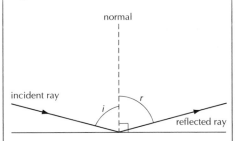

The laws of reflection are that:
- the incident angle is equal to the reflected angle.
- the incident ray, the reflected ray and the normal all lie in the same plane (as shown in the diagram).

The second statement is only included in order to be precise and is often omitted. It should be obvious that a ray arriving at a mirror (such as the one represented above) is not suddenly reflected in an odd direction (e.g. out of the plane of the page).

Formation of an image by reflection

RAY DIAGRAMS

If an object is placed in front of a plane mirror, an image will be formed.

The process is as follows:
- light sets off in all directions from every part of the object. (This is a result of diffuse reflections from a source of light.)
- each ray of light that arrives at the mirror is reflected according to the law of reflection.
- these rays can be received by any observer.
- the location of the image seen by the observer arises because the rays are assumed to have travelled in straight lines.

In order to find the location and nature of this image a ray diagram is needed.

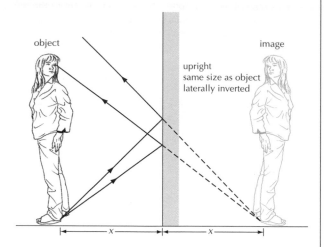

The image formed by reflection in a plane mirror is always
- **The same distance behind the mirror as the object is in front**.
- **Upright** (as opposed to being inverted).
- **The same size as the object** (as opposed to being magnified or diminished).
- **Laterally inverted** (i.e. left and right are interchanged).
- **Virtual** (see below).

REAL AND VIRTUAL IMAGES

The image formed by reflection in a plane mirror is described as a **virtual image**. This term is used to describe images created when rays of light **seem** to come from a single point but in fact they do not pass through that point. In the example above, the rays of light seem to be coming from behind the mirror. They do not, of course, actually pass behind the mirror at all.

The opposite of a virtual image is a **real image**. In this case, the rays of light do actually pass through a single point. Real images cannot be formed by plane mirrors, but they can be formed by concave mirrors or by lenses. For example, if you look into the concave surface of a spoon, you will see an image of yourself. This particular image is
- **Upside down**
- **Diminished**
- **Real**

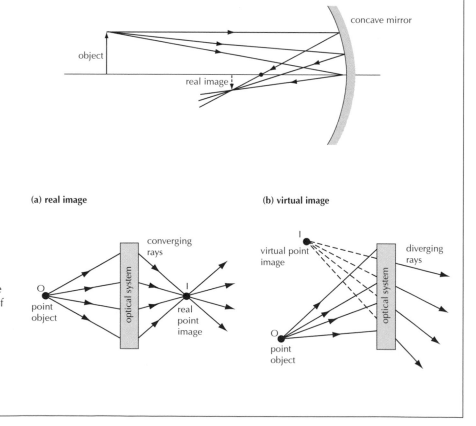

Snell's law and refractive index

REFRACTIVE INDEX AND SNELL'S LAW.

Refraction takes place at the boundary between two media. In general, a wave that crosses the boundary will undergo a change of direction. The reason for this change in direction is the change in wave speed that has taken place. See page 33 for more details.

As with reflection, the ray directions are always specified by considering the angles between the ray **and the normal.** If a ray travels into an optically denser medium (e.g. from air into water), then the ray of light is refracted **towards** the normal. If the ray travels into an optically less dense medium then the ray of light is refracted **away from** the normal.

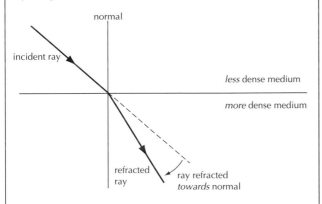

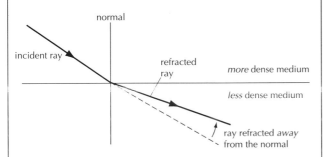

Snell's law allows us to work out the angles involved. When a ray is refracted between two different media, the ratio $\dfrac{\sin(\text{angle of incidence})}{\sin(\text{angle of refraction})}$ is a constant.

The constant is called the refractive index n between the two media. As seen on page 33 this ratio is equal to the ratio of the speeds of the waves in the two media.

$$\frac{\sin i}{\sin r} = n$$

If the refractive index for a particular substance is given as a particular number and the other medium is not mentioned then you can assume that the other medium is air (or to be absolutely correct, a vacuum). Another way of expressing this is to say that the refractive index of air can be taken to be 1.0.

For example the refractive index for a type of glass might be given as

$$n_{\text{glass}} = 1.34$$

This means that a ray entering the glass from air with an incident angle of 40° would have a refracted angle given by

$$\sin r = \frac{\sin 40}{1.34} = 0.4797$$

$$\therefore r = 28.7°$$

EXAMPLES

1. Parallel-sided block

A ray will always leave a parallel-sided block travelling in a parallel direction to the one with which it entered the block. The overall effect of the block has been to move the ray sideways. An example of this is shown below.

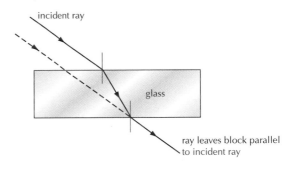

2. Ray travelling between two media

If a ray goes between two different media, the two individual refractive indices can be used to calculate the overall refraction using the following equation

$$n_1 \sin \theta_1 = n_2 \sin \theta_2$$

n_1 refractive index of medium 1

θ_1 angle in medium 1

n_2 refractive index of medium 2

θ_2 angle in medium 2

Suppose a ray of light is shone into a fish tank that contains water. The refraction that takes place would be calculated as shown below:

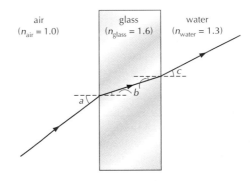

1st refraction:

$$n_{\text{glass}} = \frac{\sin a}{\sin b}$$

2nd refraction:

$$n_{\text{glass}} \times \sin b = n_{\text{water}} \times \sin c$$

$$\frac{n_{\text{glass}}}{n_{\text{water}}} = \frac{\sin c}{\sin b}$$

Overall the refraction is from incident angle a to refracted angle c.

i.e. $$n_{\text{overall}} = \frac{\sin a}{\sin c}$$

$$= n_{\text{water}}$$

Image formation

STICK IN WATER

The image formed as a result of the refraction of light leaving water is so commonly seen that most people forget that the objects are made to seem strange. A straight stick will appear bent if it is place in water. The brain assumes that the rays that arrive at one's eyes must have been travelling in a straight line.

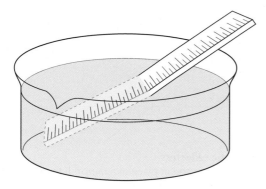

A straight stick appears bent when placed in water

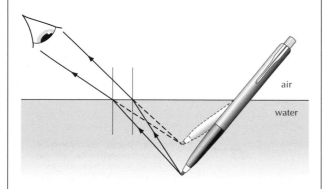

The image of the end of the pen is:
- nearer to the surface than the pen actually is.
- virtual

REAL AND APPARENT DEPTH

If one looks into a pool of water, the apparent depth of an object (including the bottom of the pool) always appears to be less than it actually turns out to be. This is because a virtual image of the object is created by the refraction that takes place at the surface of the water. If the object is being viewed from above, it is possible to work out the relationship between the real and apparent depth. The angles in the following diagram have been exaggerated in order to make the picture clearer.

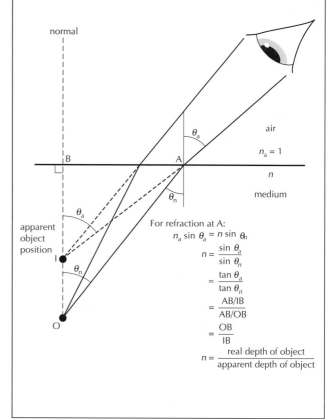

For refraction at A:

$$n_a \sin \theta_a = n \sin \theta_n$$

$$n = \frac{\sin \theta_a}{\sin \theta_n}$$

$$\approx \frac{\tan \theta_a}{\tan \theta_n}$$

$$= \frac{AB/IB}{AB/OB}$$

$$= \frac{OB}{IB}$$

$$n = \frac{\text{real depth of object}}{\text{apparent depth of object}}$$

EXAMPLE

How deep does a swimming pool appear to be given that its real depth is 3.0 m and the refractive index of water is 1.3?

Since $n = \dfrac{\text{real depth}}{\text{apparent depth}}$

apparent depth $= \dfrac{3.0}{1.3} = 2.3$ m.

Critical angle

TOTAL INTERNAL REFLECTION AND CRITICAL ANGLE

In general, both reflection and refraction can happen at the boundary between two media.

It is, under certain circumstances, possible to guarantee complete (total) reflection with no transmission at all. This can happen when a ray meets the boundary and it is travelling in the denser medium.

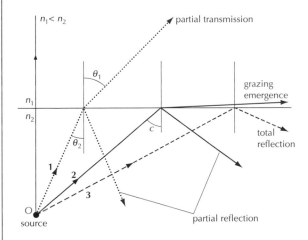

Ray1 This ray is partially reflected and partially refracted.

Ray2 This ray has a refracted angle of nearly 90°. The **critical ray** is the name given to the ray that has a refracted angle of 90°. The **critical angle** is the angle of incidence c for the critical ray.

Ray3 This ray has an angle of incidence **greater** than the critical angle. Refraction cannot occur so the ray must be totally reflected at the boundary and stay inside medium 2. The ray is said to be **totally internally reflected**.

The critical angle can be worked out as follows. For the critical ray,

$$n_1 \sin \theta_1 = n_2 \sin \theta_2$$
$$\theta_1 = 90°$$
$$\theta_2 = \theta_c$$
$$\therefore \sin \theta_c = \frac{1}{n}$$

EXAMPLES

1. What a fish sees underwater

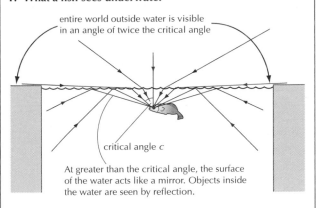

entire world outside water is visible in an angle of twice the critical angle

critical angle c

At greater than the critical angle, the surface of the water acts like a mirror. Objects inside the water are seen by reflection.

2. Prismatic reflectors

A prism can be used in place of a mirror. If the light strikes the surface of the prism at greater than the critical angle, it must be totally internally reflected.

Prisms are used in many optical devices. Examples include:

- periscopes – the double reflection allows the user to see over a crowd.
- binoculars – the double reflection means that the binoculars do not have to be too long
- SLR cameras – the view through the lens is reflected up to the eyepiece.

periscope   **binoculars**

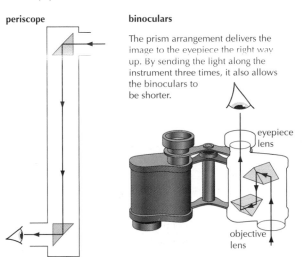

The prism arrangement delivers the image to the eyepiece the right way up. By sending the light along the instrument three times, it also allows the binoculars to be shorter.

eyepiece lens

objective lens

3. Optical fibre

Optical fibres can be used to guide light along a certain path. The idea is to make a ray of light travel along a transparent fibre by bouncing between the walls of the fibre. So long as the incident angle of the ray on the wall of the fibre is always greater than the critical angle, the ray will always remain within the fibre even if the fibre is bent.

Two important uses of optical fibres are:
- in the communication industry. Digital data can be encoded into pulses of light that can then travel along the fibres. This is used for telephone communication, cable TV etc.
- in the medical world. Bundles of optical fibres can be used to carry images back from inside the body. This instrument is called an endoscope.

Types of lens

CONVERGING LENSES

A converging lens brings parallel rays into one focus point.

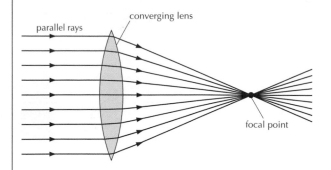

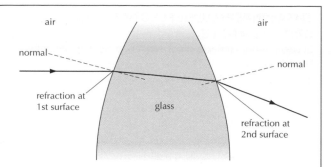

The reason that this happens is the refraction that takes place at both surfaces of the lens.

The rays of light are all brought together in one point because of the particular shape of the lens. Any one lens can be thought of as a collection of different-shaped glass blocks. It can be shown that any thin lens that has surfaces formed from sections of spheres will converge light into one focus point.

A converging lens will always be thicker at the centre when compared with the edges.

DIVERGING LENSES

A diverging lens makes parallel rays spread apart. After passing through the lens, they appear to be coming from one focus point. Once again the reason for this is the refraction that takes place at the surfaces. A diverging lens will always be thinner at the centre when compared with the edges.

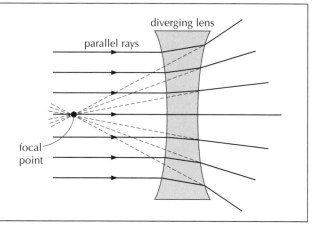

DEFINITIONS

When analysing lenses and the images that they form, some technical terms need to be defined.

- The curvature of each surface of a lens makes it part of a sphere. The **centre of curvature** for the lens surface is the centre of this sphere.
- The **principal axis** is the line going directly through the middle of the lens. Technically it joins the centres of curvature of the two surfaces
- The **focal point** (principal focus) of a lens is the point on the principal axis to which rays that were parallel to the

principal axis are brought to focus after passing through the lens. A lens will thus have a focal point on each side.
- The **focal length** is the distance between the centre of the lens and the focal point.
- The **linear magnification** is the ratio between the size (height) of the image and the size (height) of the object. It has no units.

$$\text{linear magnification} = \frac{\text{image size}}{\text{object size}}$$

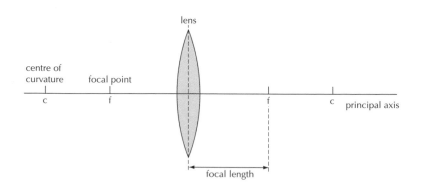

Image formation in lenses

IMPORTANT RAYS

In order to determine the nature and position of the image created of a given object, we need to construct **ray diagrams** of the set-up. In order to do this, we concentrate on the paths taken by three particular rays. As soon as the paths taken by two of these rays have been constructed, the paths of all the other rays can be inferred. These important rays are described below.

Converging lens

1. Any ray that was travelling parallel to the principal axis will be refracted towards the focal point on the other side of the lens.
2. Any ray that travelled through the focal point will be refracted parallel to the principal axis.
3. Any ray that goes through the centre of the lens will be undeviated.

Diverging lens

1. Any ray that was travelling parallel to the principal axis will be refracted away from the focal point on the same side of the lens.
2. Any ray that is heading towards the focal point on the other side of the lens will be refracted parallel to the principal axis.
3. Any ray that goes through the centre of the lens will be undeviated.

POSSIBLE SITUATIONS

A ray diagram can be constructed as follows

- an upright arrow on the principal axis represents the object.
- the paths of two important rays from the top of the object are constructed.
- this locates the position of the top of the image.
- the bottom of the image must be on the principle axis directly above (or below) the top of the image.

A full description of the image created would include the following information

- if it is real or virtual.
- if it is upright or inverted.
- if it is magnified or diminished.
- its exact position.

It should be noted that the important rays are just used to locate the image. The real image also consists of all the other rays from the object. In particular, the image will still be formed even if some of the rays are blocked off.

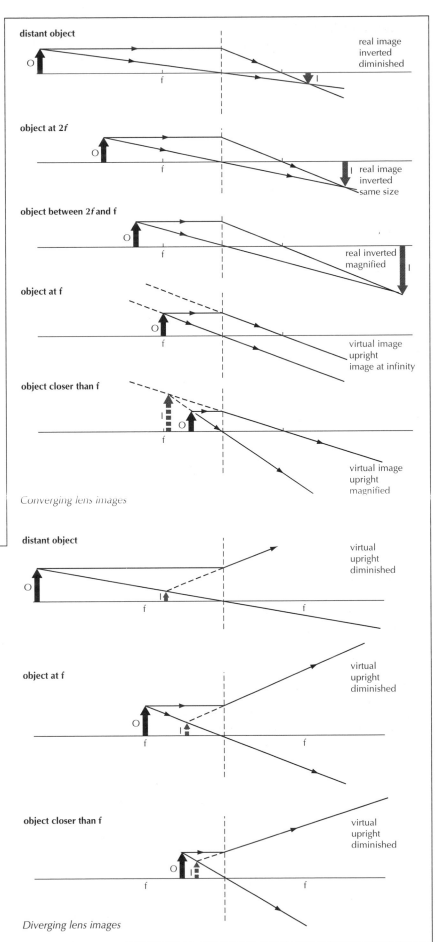

Converging lens images

Diverging lens images

Thin lens equation

LENS EQUATION

There is a mathematical method of locating the image formed by a lens.
An analysis of the angles involved shows that the following equation can be applied to thin spherical lenses:

$$\frac{1}{u} + \frac{1}{v} = \frac{1}{f}$$

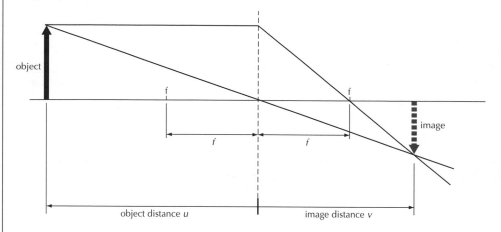

Suppose $u = 25$ cm
          $f = 10$ cm

This would mean that $\dfrac{1}{v} = \dfrac{1}{f} - \dfrac{1}{u} = \dfrac{1}{10} - \dfrac{1}{25} = \dfrac{5}{50} - \dfrac{2}{50} = \dfrac{3}{50}$

In other word, $v = \dfrac{50}{3} = 16.7$ cm

REAL IS POSITIVE

Care needs to be taken with diverging lenses or virtual images. The equation does work in these cases but for this to be the case, the following convention has to be followed:

- the focal length of diverging lens is represented by a negative value for f.
- a virtual image is represented by a negative value for u – in other words, it will be on the same side of the lens as the object.

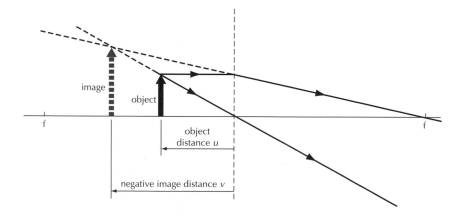

Suppose $u = 10$ cm
          $f = 25$ cm

This would mean that $\dfrac{1}{v} = \dfrac{1}{f} - \dfrac{1}{u} = \dfrac{1}{25} - \dfrac{1}{10} = \dfrac{2}{50} - \dfrac{5}{50} = -\dfrac{3}{50}$

In other word, $v = -\dfrac{50}{3} = -16.7$ cm

The simple magnifying glass

NEAR AND FAR POINT

The human eye can focus objects at different distances from the eye. Two terms are useful to describe the possible range of distances – the **near point** and the **far point**.

- The **near point** is the distance between the eye and the nearest object that can be brought into clear focus (without strain or help from, for example, lenses). It is also known as the "least distance of distinct vision". By convention it is taken to be 25 cm for normal vision.
- The **far point** is the distance between the eye and the furthest object that can be brought into focus. This is taken to be infinity for normal vision.

ANGULAR SIZE

If we bring an object closer to us (and our eyes are still able to focus on it) then we see it in more detail. This is because, as the object approaches, it occupies a bigger visual angle. The technical term for this is that the object **subtends** a larger angle.

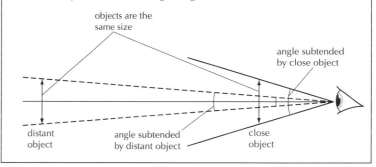

ANGULAR MAGNIFICATION

The angular magnification M of an optical instrument is defined as the ratio between the angle that an object subtends normally and the angle that its image subtends as a result of the optical instrument. The 'normal' situation depends on the context. It should be noted that the angular magnification is not the same as the linear magnification.

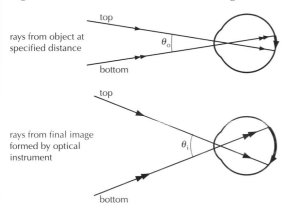

$$M = \frac{\theta_i}{\theta_o}$$

The largest visual angle that an object can occupy is when it is placed at the near point. This is often taken as the 'normal' situation.

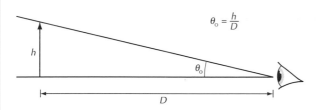

$$\theta_o = \frac{h}{D}$$

A simple lens can increase the angle subtended. It is usual to consider two possible situations.

1. Image formed at infinity

In this arrangement, the object is placed at the focal point. The resulting image will be formed at infinity and can be seen by the relaxed eye.

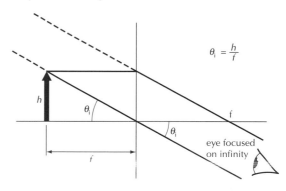

$$\theta_i = \frac{h}{f}$$

eye focused on infinity

In this case the angular magnification would be

$$M = \frac{\theta_i}{\theta_0} = \frac{h/f}{h/D} = \frac{D}{f}$$

This is the smallest value that the angular magnification can be.

2. Image formed at near point

In this arrangement, the object is placed nearer to the lens. The resulting virtual image is located at the near point. This arrangement has the largest possible angular magnification.

$$M = \frac{\theta_i}{\theta_0} = \frac{h_i/D}{h/D} = \frac{h_i}{h} = \frac{D}{a}$$

because

$$\frac{1}{u} + \frac{1}{v} = \frac{1}{f}$$

$$\frac{v}{u} = \frac{v}{f} - 1$$

here $\left|\frac{v}{u}\right| = \frac{D}{a} = \frac{D}{f} + 1$ since $v = -D$

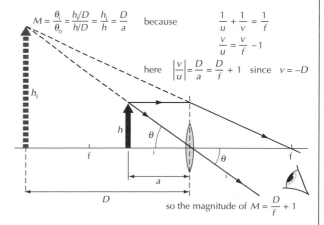

so the magnitude of $M = \frac{D}{f} + 1$

So the magnitude of $M = \frac{D}{f} + 1$

The compound microscope and astronomical telescope

COMPOUND MICROSCOPE

A compound microscope consists of two lenses – the **objective lens** and the **eyepiece lens**.
The first lens (the objective lens) forms a **real magnified** image of the object being viewed.
This real image can then be considered as the object for the second lens (the eyepiece lens)
which acts as a magnifying lens. The rays from this real image travel into the eyepiece lens
and they form a **virtual magnified** image. In normal adjustment, this virtual image is arranged
to be located at the near point so that maximum angular magnification is obtained.

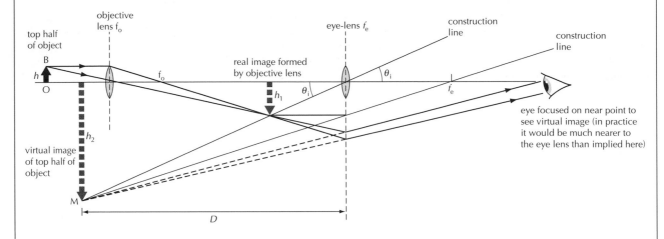

$$M = \frac{\theta_i}{\theta_o} = \frac{h_2/D}{h/D} = \frac{h_2}{h} = \frac{h_2}{h_1} \cdot \frac{h_1}{h} = \text{linear magnification produced by eye piece} \times \text{linear magnification produced by objective}$$

ASTRONOMICAL TELESCOPE

An astronomical telescope also consists of two lenses. In this case, the objective lens
forms a **real** but **diminished** image of the distant object being viewed. Once again, this
real image can then be considered as the object for the eyepiece lens acting as a
magnifying lens. The rays from this real image travel into the eyepiece lens and they
form a **virtual magnified** image. In normal adjustment, this virtual image is arranged
to be located at infinity.

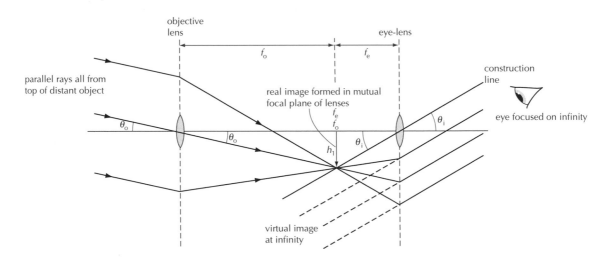

$$M = \frac{\theta_i}{\theta_o} = \frac{h_1/f_e}{h_1/f_o} = \frac{f_o}{f_e}$$

The length of the telescope $\approx f_o + f_e$.

Aberrations

SPHERICAL

A lens is said to have an **aberration** if, for some reason, a point object does not produce a perfect point image. In reality, lenses that are spherical do not produce perfect images. **Spherical aberration** is the term used to describe the fact that rays striking the outer regions of a spherical lens will be brought to a slightly different focus point from those striking the inner regions of the same lens.

In general, a point object will focus into a small circle of light, rather than a perfect point. There are several possible ways of reducing this effect:

- the shape of the lens could be altered in such a way as to correct for the effect. The lens would, of course, no longer be spherical. A particular shape only works for objects at a particular distance away.
- the effect can be reduced for a given lens by decreasing the aperture. The technical term for this is **stopping down** the aperture. The disadvantage is that the total amount of light is reduced and the effects of diffraction (see page 170) would be made worse.

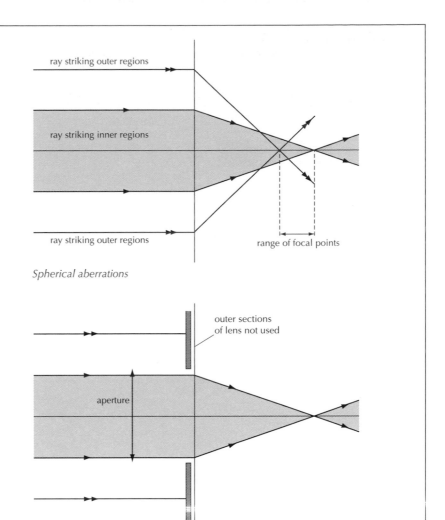

Spherical aberrations

CHROMATIC

Chromatic aberration is the term used to describe the fact that rays of different colours will be brought to a slightly different focus point by the same lens. The refractive index of the material used to make the lens is different for different frequencies of light.

The effect can be eliminated for two given colours (and reduced for all) by using two different materials to make up a compound lens. This compound lens is called an **achromatic doublet**. The two types of glass produce equal but opposite dispersion.

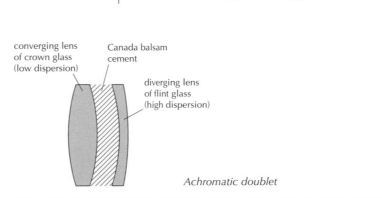

Achromatic doublet

HL Diffraction

BASIC OBSERVATIONS

Diffraction is a wave effect. The objects involved (slits, apertures etc.) have a size that is of the same order of magnitude as the wavelength of visible light.

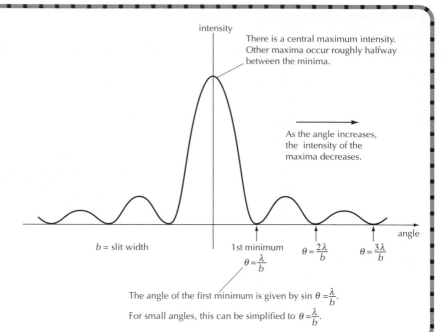

Nature of obstacle	Geometrical shadow	Diffraction pattern
(a) straight edge		
(b) single long slit $b \sim 3\lambda$		
(c) circular aperture		
(d) single long slit $b \sim 5\lambda$		

The intensity plot for a single slit is shown right.

intensity

There is a central maximum intensity. Other maxima occur roughly halfway between the minima.

As the angle increases, the intensity of the maxima decreases.

b = slit width

1st minimum $\theta = \frac{\lambda}{b}$

$\theta = \frac{2\lambda}{b}$

$\theta = \frac{3\lambda}{b}$

angle

The angle of the first minimum is given by $\sin \theta = \frac{\lambda}{b}$.

For small angles, this can be simplified to $\theta = \frac{\lambda}{b}$.

EXPLANATION

The shape of the relative intensity versus angle plot can be derived using Huygens' principle. We can treat the slit as a series of secondary wave sources. In the forward direction (θ = zero) these are all in phase so they add up to give a maximum intensity. At any other angle, there is a path difference between the rays that depends on the angle.

The overall result is the addition of all the sources. The condition for the first minimum is that the angle must make all of the sources across the slit cancel out.

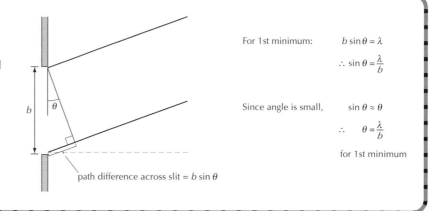

path difference across slit = $b \sin \theta$

For 1st minimum: $b \sin \theta = \lambda$

$\therefore \sin \theta = \frac{\lambda}{b}$

Since angle is small, $\sin \theta \approx \theta$

$\therefore \quad \theta = \frac{\lambda}{b}$

for 1st minimum

DOUBLE SLIT INTERFERENCE

The double slit interference pattern shown on page 74 was derived assuming that each slit was behaving like a perfect point source. This can only take place if the slits are infinitely small. In practice they have a finite width. The diffraction pattern of each slit needs to be taken into account when working out the overall double slit interference pattern as shown below.

Decreasing the slit width will mean that the observed pattern becomes more and more 'idealised'. Unfortunately, it will also mean that the total intensity of light will be decreased. The interference pattern will become harder to observe.

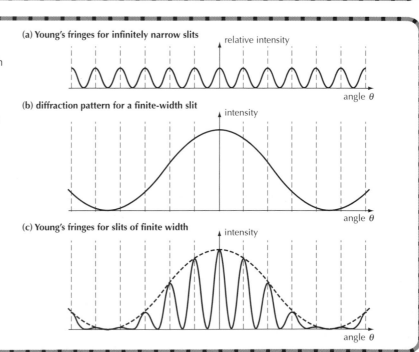

(a) Young's fringes for infinitely narrow slits

relative intensity

angle θ

(b) diffraction pattern for a finite-width slit

intensity

angle θ

(c) Young's fringes for slits of finite width

intensity

angle θ

HL Resolution

DIFFRACTION AND RESOLUTION

If two sources of light are very close in angle to one another, then they are seen as one single source of light. If the eye can tell the two sources apart, then the sources are said to be **resolved**. The diffraction pattern that takes place at apertures affects the eye's ability to resolve sources. The examples to the right show how the appearance of two line sources will depend on the diffraction that takes place at a slit. The resulting appearance is the addition of the two overlapping diffraction patterns. The graph of the resultant relative intensity of light at different angles is also shown.

These examples look at the situation of a line source of light and the diffraction that takes place at a slit. A more common situation would be a point source of light, and the diffraction that takes place at a circular aperture. The situation is exactly the same, but diffraction takes place all the way around the aperture. As seen on the previous page, the diffraction pattern of the point source is thus concentric circles around the central position. The geometry of the situation results in a slightly different value for the first minimum of the diffraction pattern.

For a **slit**, the first minimum was at the angle

$$\theta = \frac{\lambda}{b}$$

For a **circular aperture**, the first minimum is at the angle

$$\theta = \frac{1.22\,\lambda}{b}$$

If two sources are just resolved, then the first minimum of one diffraction pattern is located on top of the maximum of the other diffraction pattern. This is known as the **Rayleigh criterion**.

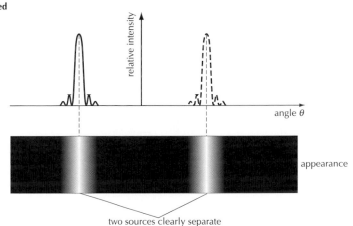

(a) resolved

two sources clearly separate

appearance

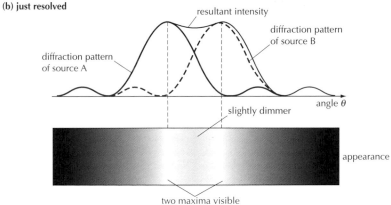

(b) just resolved

resultant intensity

diffraction pattern of source A

diffraction pattern of source B

slightly dimmer

appearance

two maxima visible

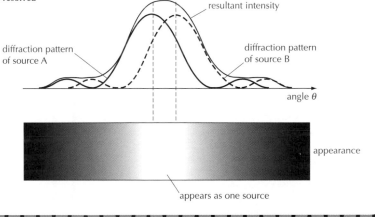

(c) not resolved

resultant intensity

diffraction pattern of source A

diffraction pattern of source B

angle θ

appearance

appears as one source

EXAMPLE

Late one night, a student was observing a car approaching from a long distance away. She noticed that when she first observed the headlights of the car, they appeared to be one point of light. Later, when the car was closer, she became able to see two separate points of light. If the wavelength of the light can be taken as 500 mm and the diameter of her pupil is approximately 4 mm,

calculate how far away the car was when she could first distinguish two points of light. Take the distance between the headlights to be 1.8 m.

When just resolved

$$\theta = \frac{1.22 \times \lambda}{b}$$

$$= \frac{1.22 \times 5 \times 10^{-7}}{0.004}$$

$$= 1.525 \times 10^{-4}$$

Since θ small

$$\theta = \frac{1.8}{x} \quad [x \text{ is distance to car}]$$

$$\Rightarrow \quad x = \frac{1.8}{1.525 \times 10^{-4}}$$

$$= 11.803$$

$$\simeq 12 \text{ km}$$

Optics 171

Multiple slit diffraction

THE DIFFRACTION GRATING

The diffraction that takes place at an individual slit affects the overall appearance of the fringes in Young's double slit experiment (see page 170 for more details). This section considers the effect on the final interference pattern of adding further slits. A series of parallel slits (at a regular separation) is called a **diffraction grating**.

Additional slits at the same separation will not effect the condition for constructive interference. In other words, the angle at which the light from slits adds constructively will be unaffected by the number of slits. The situation is shown below.

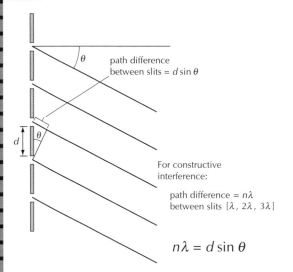

path difference
between slits = $d \sin \theta$

For constructive interference:

path difference = $n\lambda$
between slits [λ, 2λ, 3λ]

$$n\lambda = d \sin \theta$$

This formula also applies to the Young's double slit arrangement. The difference between the patterns is most noticeable at the angles where perfect constructive interference does not take place. If there are only two slits, the maxima will have a significant angular width. Two sources that are just out of phase interfere to give a resultant that is nearly the same amplitude as two sources that are exactly in phase.

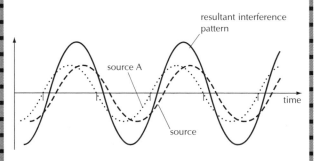

resultant interference pattern

source A

source

The addition of more slits will mean that each new slit is just out of phase with its neighbour. The overall interference pattern will be totally destructive.

overall interference pattern is totally destructive

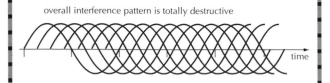

The addition of further slits at the same slit separation has the following effects:
- the principal maxima maintain the same separation.
- the principal maxima become much sharper.
- the overall amount of light being let through is increased, so the pattern increases in intensity.

Of course this pattern is still modified by the diffraction that takes place at each slit in a way that is similar to the modification introduced on page 170.

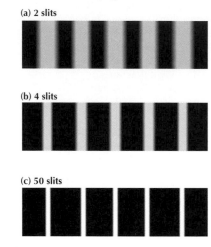

(a) 2 slits

(b) 4 slits

(c) 50 slits

Grating patterns

USES

One of the main uses of a diffraction grating is the accurate experimental measurement of the different wavelengths of light contained in a given spectrum. If white light is incident on a diffraction grating, the angle at which constructive interference takes place depends on wavelength. Different wavelengths can thus be observed at different angles. The accurate measurement of the angle provides the experimenter with an accurate measurement of the exact wavelength (and thus frequency) of the colour of light that is being considered. The apparatus that is used to achieve this accurate measurement is called a **spectrometer**.

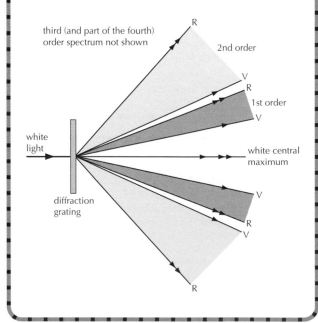

third (and part of the fourth) order spectrum not shown

2nd order

1st order

white light

white central maximum

diffraction grating

HL Parallel films

PHASE CHANGES

There are many situations when interference can take place that also involve the reflection of light. When analysing in detail the conditions for constructive or destructive interference, one needs to take any **phase changes** into consideration. A phase change is the inversion of the wave that can take place at a reflection interface, but it does not always happen. It depends on the two media involved.

The technical term for the inversion of a wave is that it has 'undergone a phase change of π'.

- When light is reflected back from an optically denser medium there is a phase change of π.
- When light is reflected back from an optically less dense medium there is no phase change.

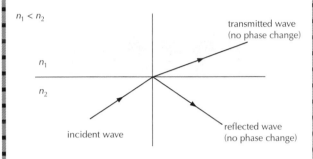

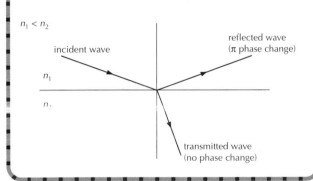

CONDITIONS FOR INTERFERENCE PATTERNS

A parallel-sided film can produce interference as a result of the reflections that are taking place at both surfaces.

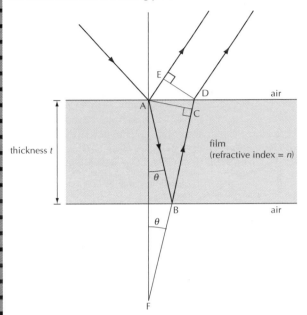

From point A, there are two possible paths:

1 along path AE in air
2 along ABCD in film

These rays then interfere and we need to calculate the optical path difference.

The path AE in air is equivalent to CD in the film.
So path difference = (AB + BC) in film.

In addition, the phase change at A is equivalent to $\frac{\lambda}{2}$ path difference.

So total path difference = (AB + BC) in film + $\frac{\lambda}{2}$

$$= n(AB + BC) + \frac{\lambda}{2}$$

By geometry:

$$(AB + BC) = FC$$
$$= 2t \cos \theta$$

∴ path difference $= 2nt \cos \theta + \frac{\lambda}{2}$

if path difference $= n\lambda$: constructive

if path difference $= \left(n + \frac{1}{2}\right)\lambda$: destructive

EXAMPLE

The above equations work out the angles for which constructive and destructive interference take place for a given wavelength. If the source of light is an extended source, the eye receives rays leaving the film over a range of values for θ.

If white light is used then the situation becomes more complex. Provided the thickness of the film is small, then one or two colours may reinforce along a direction in which others cancel. The appearance of the film will be bright colours, such as can be seen when looking at

- an oil film on the surface of water or
- soap bubbles.

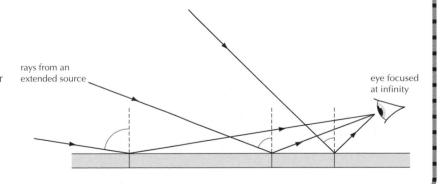

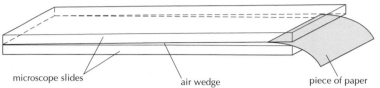 Wedge films

THEORY

If two glass plates are at a small angle to one another, the gap in-between the two is called an air wedge. An example of this set-up would be two microscope slides with a piece of paper between them at one end.

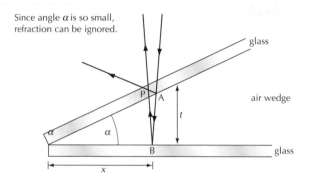

microscope slides   air wedge   piece of paper

There is a path difference between the rays of light reflecting from the top and from the bottom surfaces of the air wedge. This results in parallel lines of equally spaced constructive and destructive interference fringes.

Since angle α is so small, refraction can be ignored.

glass

air wedge

P A

t

α

α

B   glass

x

$\therefore$ path difference due to air wedge $\approx 2t$.
A phase change happens at B, but not A.

$\therefore$ path difference $= 2t + \dfrac{\lambda}{2}$

$\therefore$ for a bright fringe at P:

$$2t + \frac{\lambda}{2} = n\lambda$$

Since $\quad\quad\quad t = \alpha x$

then $\quad\quad\quad x = \left(n - \dfrac{1}{2}\right)\dfrac{\lambda}{2\alpha}$

so fringe separation $= \dfrac{\lambda}{2\alpha}$

Exaggerated diagram of air wedge

equally spaced fringes

USE

If the separation of the fringes is measured and the wavelength of light used is known, the angle and thus the thickness of the wedge at any point can be calculated. This method can be used for measuring very small distances. Any small change in distance will result in a movement of the fringe pattern. Using this technique, a distance change of the order of about 10^{-7} m can be recorded.

IB QUESTIONS – OPTION H – OPTICS

1 A student is given two converging lenses, A and B, and a tube in order to make a telescope.

   (a) Describe a simple method by which she can determine the focal length of each lens. [2]

   (b) She finds the focal lengths to be as follows:

      Focal length of lens A   10 cm
      Focal length of lens B   50 cm

      Draw a diagram to show how the lenses should be arranged in the tube in order to make a telescope. Your diagram should include:

      (i) labels for each lens;
      (ii) the focal points for each lens;
      (iii) the position of the eye when the telescope is in use. [4]

   (c) On your diagram, mark the location of the intermediate image formed in the tube. [1]

   (d) Is the image seen through the telescope upright or upside-down? [1]

   (e) Approximately how long must the telescope tube be? [1]

2 An elderly lady buys a 'magnifying glass' to read small print in the telephone directory. To her surprise she finds that if she holds the convex lens fairly close to the page she gets one kind of image, while if she holds it fairly far from the page she gets quite another kind of image.

   (a) *Lens close to the page.* For the lens quite close to the page she draws the ray diagram below.

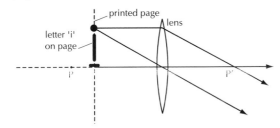

      (i) Where should her eye be located in order to see the image of the letter 'i'? Tick the correct answer below. [1]

      To the left of the lens   ☐
      Anywhere to the right of the lens ☐
      To the right of the focal point P' ☐

      (ii) If she looks at letters on a page in this way, how will they appear to her?

      Right way up or upside down?
      Enlarged or diminished?
      Behind the lens or in front of the lens?
      Should she be able to read the telephone directory using the lens this way? [2]

   (b) *Lens further from the page.* The lady now moves the page further from the lens. The diagram below represents the situation where the page is more than twice the focal distance from the lens.

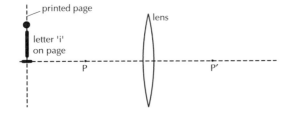

      (i) Locate the image of the 'i' by tracing suitable rays on the diagram. [4]

      (ii) Where should the lady's eye be located in order to see the image? Tick the correct answer below. [1]

      To the left of the lens   ☐
      Anywhere to the right of the lens ☐
      Between the lens and P'   ☐
      To the right of the image   ☐

      (iii) If she looks at letters on a page in this way, how will they appear to her?

      Right way up or upside-down?
      Enlarged or diminished?
      Behind the lens or in front of the lens?
      Nearer or further away than the page?

      Will she be able to read the telephone directory using the lens this way? [3]

3 Abigail looks at a particular star with her naked eye and she sees the star as a point of light. When she looks at the star through a telescope she sees that there are two points of light. The star Abigail is looking at is actually two stars close together.

   (a) Explain, assuming that Abigail's eyes are functioning normally,
      (i) why she is unable to distinguish the two stars with her naked eye. [3]
      (iii) how the telescope enables her to distinguish the two stars. [2]

   (b) The system that Abigail is observing is 4.2×10^{16} m from the Earth and the two stars are separated by a distance of 2.6×10^{11} m. Assuming that the average wavelength of the light emitted by the stars is 500 nm, estimate the minimum diameter of the objective lens of a telescope that will just enable the two stars to be distinguished. [3]

4 This question is about diffraction.

In the diagram below (not to scale) a monochromic beam of light of wavelength 500 nm is incident on a single slit of width 0.1 mm. After passing through the slit the light is brought to a focus (the focusing lens is not shown) on a screen placed at a distance of 1.0 m from the slit.

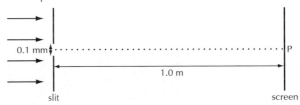

   (a) On the axes below sketch a diagram showing how the intensity of the light varies at different points along the screen. (Note that this is a sketch graph; no values are required.) [3]

      intensity
      P   distance along screen

   (b) Calculate the width of the central maximum. [3]

Answers to questions

MEASUREMENT *(page 8)*
1 B 2 B 3 B 4 C 5 A 6 B 7 B 8 C
9 **(a) (i)** $0.5 \times$ acceleration down slope **(iv)** 0.36 m s^{-2}

MECHANICS *(page 22)*
1 C 2 D 3 B 4 B 5 C
6 **(a)** equal **(b)** left **(c)** 20 km h^{-1} **(e)** Car driver **(f)** No
7 **(a)** Yes, in towards the centre of the circle
  (c) 9.6 m s^{-1}

THERMAL PHYSICS *(page 30)*
1 B 2 D 3 D 4 D
5 **(a) (i)** Length = 20m; depth = 2m; width = 5m; temp = 25°C
    (ii) $464
  (b) (i) 84 days

WAVES *(page 38)*
1 C 2 C 3 C 4 C
5 **(b)** 1200 m s^{-1} **(d)** 10 kHz **(e)** 0.12 m **(f)** none
6 **(b)** 0.8 m
  (c) 2 or 3 × fundamental (depending on answer given to (a))
  (d) 6.4 cm from end

ELECTRICITY AND MAGNETISM *(page 47)*
1 D 2 A 3 A 4 B 5 D
6 **(b)** equal **(c)** A & C brighter; B dimmer **(d)** 1.08 W
7 **(a)** same **(b)** 50 A **(c)** increasing acceleration

ATOMIC AND NUCLEAR PHYSICS *(page 55)*
1 B 2 D 3 A 4 D 5 B 6 C 7 D
8 **(b) (i)** $^{2}_{1}\text{H} + ^{26}_{12}\text{Mg} \rightarrow ^{24}_{11}\text{Na} + ^{4}_{2}\text{He}$
9 **(a)** $^{12}_{6}\text{C} \rightarrow ^{12}_{7}\text{N} + ^{0}_{-1}\beta$ **(b) (ii)** 11 600 years
10 **(a) (i)** 3 **(b) (i)** 1.72×10^{19}

MEASUREMENT AND UNCERTAINTIES *(page 59)*
1 C 2 D 3 D 4 A
5 **(b) (i)** −3 **(ii)** 2.6×10^{-4} N m^{-3}
6 **(b)** 2.4 ± 0.1 s **(c)** 2.6 ± 0.2 m s^{-2}

HL MECHANICS (SL OPTION A) *(page 66)*
1 B 2 B 3 D 4 B 5 C
6 **(a) (i)** 1.9×10^{11} J **(ii)** 7.7 km s^{-1} **(iii)** 2.2×10^{12} J
  (c) 2.6 h
7 **(b) (i)** 3.9 m s^{-1}

THERMAL PHYSICS *(page 71)*
1 D 2 A 3 C 4 A
5 **(a)** No **(b)** equal **(c)** 300 J **(d)** −500 J **(e)** 500 J **(f)** 150 J
  (g) 16%
6 **(b)** 990 K **(c) (i)** 1 **(ii)** 2 and 3 **(iii)** 3

WAVE PHENOMENA *(page 75)*
1 B 2 C 3 D 4 C 5 B 6 C 7 C
8 **(b)** 27.5 m s^{-1}
9 **(c) (i)** A=1000 Hz; B=1100 Hz **(ii)** 100 Hz **(d) (ii)** Smaller

ELECTROMAGNETISM *(page 80)*
1 D 2 A 3 D 4 C 5 B 6 D 7 D
8 **(b)** 0.7 V

**QUANTUM PHYSICS AND NUCLEAR PHYSICS
(SL OPTION B)** *(page 88)*
1 D 2 C 3 A 4 B 5 D
6 **(b)** ln R and t **(c)** Yes **(e)** 0.375 h^{-1} **(h)** 1.85 h
8 **(a)** UDD
  (b) (i) weak interaction **(ii)** Q = −1; B = 0; L = 0 **(iii)** D→U

OPTION C – ENERGY EXTENSION *(page 94)*
1 **(c)** 15 MW **(d) (i)** 20%
4 **(a)** 1000 MW **(b)** 1200 MW **(c)** 17% **(d)** 43 kg s^{-1}
5 **(c)** 1.8 MW

OPTION D – BIOMEDICAL PHYSICS *(page 108)*
1 **(b)** 1500 Hz **(c)** 200 → 4300 Hz
2 **(b)** 3.4 Nm **(c)** 68 N
4 **(c)** 10^{-8} W m^{-2}
5 **(c)** $S \sin (70) \times 0.8$
6 **(b)** 0.8%

**OPTION E – THE HISTORY AND DEVELOPMENT OF
PHYSICS** *(page 122)*
2 **(c)** 2×10^{-10} m
3 **(d) (i)** 2.70×10^{-3} m s^{-2} **(ii)** 2.72×10^{-3} m s^{-2} **(iii)** Equal

OPTION F – ASTROPHYSICS *(page 140)*
1 **(a)** Aldebaran **(c)** Aldebaran **(e)** Smaller **(f)** − 2.9 **(g)** 200 pc
2 **(a)** 5800 K
3 **(e)** 499.83 nm

OPTION G – SPECIAL AND GENERAL RELATIVITY
(page 156)
1 **(c)** Front **(d)** T: 100 m, S: 87 m **(e)** T: 75 m, S: 87 m
2 **(b)** 31 m **(c)** 1.1×10^{-7} s **(d)** 2.9×10^{-30} kg
3 **(a) (i)** zero **(ii)** $2.7 \, m_0 c^2$
  (b) (i) $0.923 \, c$ **(ii)** $2.4 \, m_0 c$ **(iii)** $3.6 \, m_0 c^2$
  (c) agree
4 **(c)** no **(d)** Anna

OPTION H – OPTICS *(page 175)*
1 **(d)** upside-down **(e)** 60 cm
2 **(a) (i)** Anywhere to the right of the lens
      (ii) right way up; enlarged; behind the lens; yes
  (b) (ii) to the right of the image
      (iii) upside-down; diminished; in front; nearer; no
3 **(b)** 10 cm
4 **(b)** 10 mm

Origin of individual questions

The questions detailed below are all taken from past IB examination papers and are © IBO. The questions are from the May (M) or November (N), 1998 (98), 1999 (99), 2000 (00), and 2001 (01), Specimen (Sp), Standard level (S1/2/3) or Higher level (H1/2/3) papers with the question number in brackets.

MEASUREMENT
1 M99SpSI(1) **2** M99S1(1) **3** N99S1(1) **4** M00S1(1)
5 M00S1(2) **6** N00S1(1) **7** N01S1(1) **8** N01S1(2)
9 N99S2(S2)

MECHANICS
1 M98S1(2) **2** M98S1(4) **3** M98S1(8) **4** M98S1(9)
5 N98S1(4) **6** N00H2(B2) **7** N00H2(B1)

THERMAL PHYSICS
1 N99H1(15) **2** N99H1(16) **3** N99H1(17)
5 M99SpS2(B3) **6** M99S2(B3)

WAVES
1 M01H1(14) **2** M00H1(19) **3** M98H1(24)
4 M98S2(B2) **5** N99H2(B2) **6** N99H2(B1)

ELECTRICITY AND MAGNETISM
1 M99H1(30) **2** M98H1(29) **3** M98H1(30)
4 N99SpH1(23) **5** N99H1(27) **6** M99H2(A3)
7 N00S2(A3)

ATOMIC AND NUCLEAR PHYSICS
1 N98S1(29) **2** M99S1(29) **3** M99S1(30) **4** M98SpS1(29)
5 M98SpS1(30) **6** M98S1(29) **7** M98S1(30)
8 M98S2(A3) **9** M99S2(A3) **10** M99H2(B4)

MEASUREMENT AND UNCERTAINTIES
1 M98H1(5) **2** N98H1(5) **3** M99H1(3) **4** M99H1(4)
5 N98H2(A1) **6** M98SpH2(A2)

MECHANICS (SL OPTION A)
1 M99H1(10) **2** M00H1(9) **3** N00H1(7) **4** M01H1(13)
5 M01H1(1) **6** N98H2(B4) **7** N01H2(A3)

THERMAL PHYSICS (SL OPTION C)
1 M99H1(18) **2** M00H1(16) **3** M00H1(18)
4 M01H1(25) **5** N01H2(B1) **6** N98H2(A2)

WAVE PHENOMENA
1 M00H1(24) **2** M99H1(22) **3** M99H1(23) **4** N01H1(22)
5 N01H1(24) **6** N01H1(25) **7** N00H1(24) **8** N98H2(A5)
9 N00H2(B1)

ELECTROMAGNETISM
1 N00H1(31) **2** N99H1(28) **3** N99H1(29) **4** M99H1(32)
5 M98H1(33) **6** M99H1(27) **7** N99H1(34) **8** N98H2(A4)

QUANTUM AND NUCLEAR PHYSICS (SL OPTION B)
1 N98H1(32) **2** M98H1(39) **3** M99H1(40) **4** M01H1(35)
5 M01H1(36) **6** N00H2(A1) **7** M98H3(E4) **8** N99H3(E3)

OPTION C – ENERGY EXTENSION
1 N01S3(C1) **2** M99S3(C1) **3** M98SpS3(C3)
4 M98SpS3(C2) **5** M98S3(C2) **6** N98S3(C2)

OPTION D – BIOMEDICAL PHYSICS
1 N01S3(D1) **2** M98S3(D3) **3** M99H3(D2)
4 M99H3(D3) **5** N01H3(D2) **6** N01H3(D3)

OPTION E – THE HISTORY AND DEVELOPMENT OF PHYSICS
1 M00H3(E2) **2** N00H3(E1) **3** N99H3(E1) **4** N98H3(E4)

OPTION F – ASTROPHYSICS
1 N01H3(F1) **2** N01H3(F2) **3** N98H3(F2) **4** N00H3(F2)

OPTION G – SPECIAL AND GENERAL RELATIVITY
1 M00H3(G1) **2** N00H3(G2) **3** N01H3(G2)
4 M00H3(G2)

OPTION H – OPTICS
1 N00H3(H1) **2** M00H3(H1) **3** M01H3(H4)
4 M01H3(H2)

Index